have

WILD SEA

For my family

WILD SEA

A HISTORY OF THE SOUTHERN OCEAN

JOY McCANN

THE UNIVERSITY OF CHICAGO PRESS
CHICAGO AND LONDON

The University of Chicago Press, Chicago 60637
The University of Chicago Press, Ltd., London

Published 2019
Printed in the United States of America

28 27 26 25 24 23 22 21 20 19   1 2 3 4 5

ISBN-13: 978-0-226-62238-5 (cloth)
ISBN-13: 978-0-226-62241-5 (e-book)
DOI: https://doi.org/10.7208/chicago/9780226622415.001.0001

An earlier version of the work was first published in Australia by NewSouth, an imprint of UNSW Press Ltd., 2018.

LIBRARY OF CONGRESS CATALOGING-IN-PUBLICATION DATA
Names: McCann, Joy, 1954– author.
Title: Wild sea : a history of the Southern Ocean / Joy McCann.
Description: Chicago ; London : The University of Chicago Press, 2019. | "An earlier version of this work was first published in Australia by NewSouth, an imprint of UNSW Press Ltd., 2018."—Title page verso. | Includes bibliographical references and index.
Identifiers: LCCN 2018043684 | ISBN 9780226622385 (cloth : alk. paper) | ISBN 9780226622415 (ebook)
Subjects: LCSH: Antarctic Ocean. | Antarctic Ocean—Discovery and exploration.
Classification: LCC GC461 .M33 2019 | DDC 910.9167—dc23
LC record available at https://lccn.loc.gov/2018043684

♾ This paper meets the requirements of ANSI/NISO Z39.48-1992 (Permanence of Paper).

CONTENTS

The Southern Ocean surrounds Antarctica, but its northern limits have eluded precise definition and remain contested

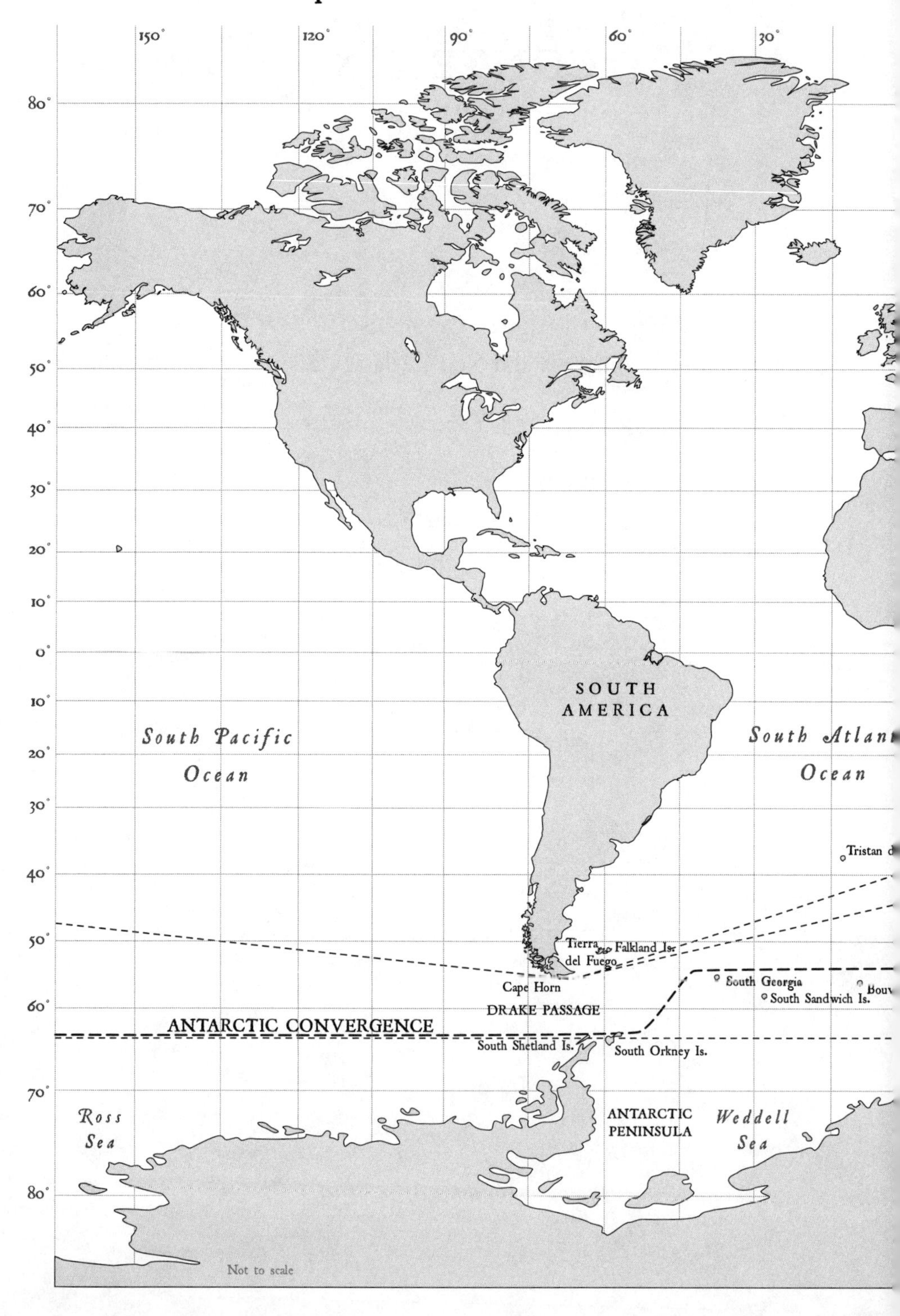

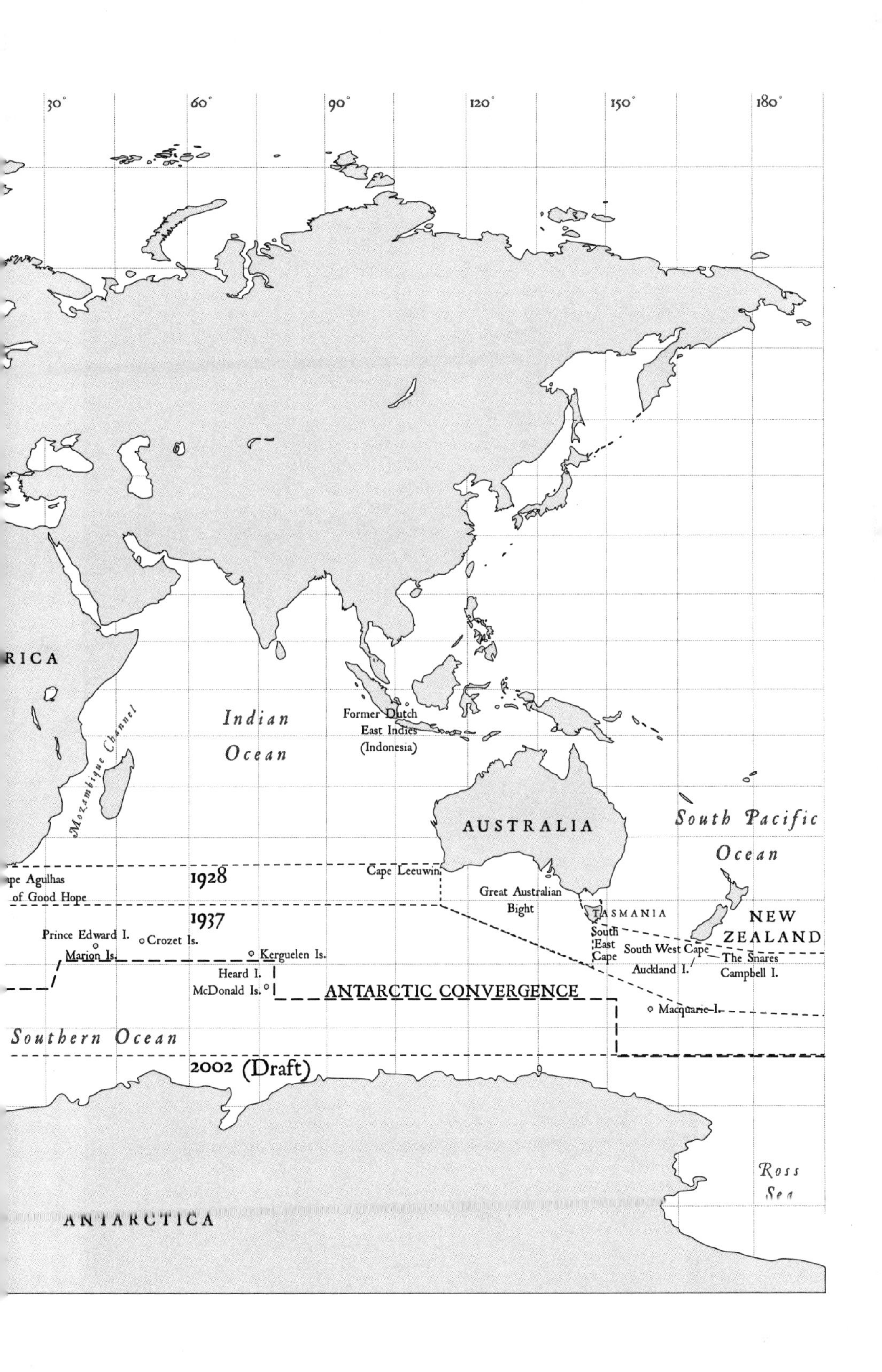

30°
60°
90°
120°
150°
180°
RICA
Mozambique Channel
Indian Ocean
Former Dutch East Indies (Indonesia)
AUSTRALIA
South Pacific Ocean
1928
Cape Leeuwin
Great Australian Bight
TASMANIA
NEW ZEALAND
1937
South East Cape
South West Cape
The Snares
Auckland I.
Campbell I.
Prince Edward I.
Crozet Is.
Marion Is.
Kerguelen Is.
Heard I.
McDonald Is.
ANTARCTIC CONVERGENCE
Macquarie I.
Southern Ocean
2002 (Draft)
Ross Sea
ANTARCTICA

PRELUDE

It is not possible to measure the full extent of that sea except with the eye of fantasy. No one will ever delve to the bottom of that sea except by plunging into the waves of his wildest dreams.

Muḥammad Rabī' ibn Muḥammad Ibrāhīm,
The Ship of Sulaimān, 1685[1]

The Southern Ocean is a wild and elusive place, an ocean like no other. With its waters lying between the Antarctic continent and the southern coastlines of Australia, New Zealand, South America and South Africa, it is the most remote and inaccessible part of the planetary ocean, the only part that flows completely around Earth unimpeded by any landmass. It is notorious amongst sailors for its tempestuous winds and hazardous fog and ice. Yet it is a difficult ocean to pin down. Its southern boundary, defined by the icy continent of Antarctica, is constantly moving in a seasonal dance of freeze and thaw. To the north, with no continental landmasses to interrupt their flow, its waters meet and mingle with those of the Atlantic, Indian and Pacific oceans along a fluid boundary that defies the neat lines of a cartographer.[2] As Elisabeth Mann Borgese, an international expert in maritime law, once observed, 'Fish, currents, waves and winds respect no "boundaries" contrived by human minds.'[3] Even the ocean's name is uncertain; it has been known by many: Antarctic Ocean, Antarctic Circumpolar Ocean, Great Southern Ocean, Southern Icy Ocean, Grand Ocean, South Polar Ocean, Austral Ocean and simply the South Atlantic, South Indian and South Pacific oceans. I have chosen to use Southern Ocean in the following pages, based on the common acceptance of the term in the Southern Hemisphere.

My earliest memories are of the Southern Ocean. I remember learning to swim in its shallows on the South Australian coast near Adelaide during the 1960s, imagining that the huge waves that crashed onto the long white beaches had travelled all the way from Antarctica. I would strain my eyes to the horizon, picturing floating ice on those sweltering Adelaide summer days. I was never much of a swimmer, but the Southern Ocean always held a peculiar fascination for me. My family had crossed that ocean to Australia on an ageing migrant ship. Much later, as a historian interested in Australian landscapes and environments, I made the unsettling discovery that those responsible for defining the boundaries of oceans and seas had erased the Southern Ocean from world maps sometime before I was born. It seems that no one thought to tell the good people of Adelaide, since its waters still surged onto their local beaches, bringing gleaming ribbons of kelp and other riches from its depths. I was intrigued, and so began my own journey into that wild ocean. In the following pages I continue that voyage, navigating back and forth not only across the physical ocean but also through its history, and through humankind's shifting political, scientific and cultural relationships with it.

In Western cultures people have found a myriad of ways to create meaning and order in the oceanic realm. Atlantic (from the Greek Atlantikos, referring to Atlas, the Titan of Greek mythology) originally described the Atlas Mountains, in North Africa, and the area of ocean off the west African coast. Atlantic Ocean later came to mean the whole oceanic region from Europe and west Africa to the Americas. The Indian subcontinent gave its name to the ocean that washed its shores, while the name Pacific Ocean (from the Portuguese Mar Pacífico, meaning Pacific or Peaceful Sea) came into European history via the Portuguese explorer Ferdinand Magellan, who described it in 1520 during a Spanish expedition to the East Indies (South Asia) in 1519–22. The land-locked Arctic Ocean (from the Greek *arktos*, meaning bear, in reference to the northern constellation of stars known as the Great Bear) received little attention from Europeans until the twentieth century. One of the earliest references to a Southern Ocean appeared in the English theologian and historian Peter Heylyn's influential *Cosmographie* first published in 1652, in which he endeavoured to describe every aspect of the known world of his century.[4]

When European explorers navigated their way into the uncharted waters of the high southern latitudes in the eighteenth century in search of new territory, resources and geographical knowledge, they encountered vast barriers of sea ice and strange winds and currents. Such early expeditions were motivated by the prospect of finding new trade routes. The British navigator James Cook was instructed by the British Admiralty to undertake three voyages of exploration to the region. He did so between 1768 and 1779, first to the Pacific Ocean to observe the transit of Venus and search for a continent thought to lie across the South Pacific below latitude 40° South. On his second voyage he crossed the Antarctic Circle for the first time but saw only ice.

Over the following two centuries larger sailing vessels and improved navigation techniques ushered in a new era of long-distance voyaging to the Southern Hemisphere. The ocean at the southernmost end of Earth began to take shape on nautical maps and charts, but sailing over such vast distances was not for the faint-hearted and shipwrecks were common. The idea of a Great Southern Land persisted until mariners venturing into the high southern latitudes mapped two great lands of desert and ice separated by a stormy, tempestuous ocean. Along the way they found rich whaling grounds in the Antarctic Circumpolar Current, where the southern whales migrated along ancestral pathways.

The prospect of a new frontier at the South Pole fuelled epic voyages of science and exploration, and nations sought to impose order on the wild ocean that surrounded it by mapping its surface features and condensing the ceaseless motions of current and wind and ice into lines on charts. Maritime explorers, natural philosophers and scientists also sought to unravel the Southern Ocean's mysteries. Living organisms were captured and preserved in the archives of natural history museums, the artworks of galleries and the records of research institutions. A combination of developments in undersea surveillance technology after World War II, which made new tools and methods available to the oceanographic and biological sciences, together with the decline of whaling and rise of a new ecological consciousness opened windows to the deep sea and created the conditions for the transformation of the stormy moat encircling Antarctica into a global field laboratory. Scientists examined the interplay of ocean and atmosphere, Earth's two great bodies of water. Satellite and sonar technologies plotted more accurately the shifting contours of current and air and mapped the invisible pathways linking this polar region to the rest of the planet. Over the centuries, each voyage, each chart, each satellite image added another fragment to Western knowledge, a process that continues to this day.

Meanwhile, the language of the Southern Ocean is saturated with Western narratives of heroic voyages and the search for new resources and knowledge. Its human archives of maps, journals and reports by sea captains, sailors, explorers and scientists are brimming with words that evoke its potential as an unlimited resource. Its whales and seals, birds and fish have become 'stock' to be harvested or managed, their value measured in terms of barrels or biomass. The legend that the Southern Ocean – like all oceans – is timeless and unchanging continues to prevail, obscuring the full extent of human impacts. Western perceptions of the Southern Ocean have also served to erase or marginalise traditional stories and local connections with the winds and waters and creatures of the high southern latitudes.[5] As human knowledge of the physical and biological nature of the Southern Ocean continues to evolve, so too does the story of people's engagement – real or imagined – with this remote and tempestuous realm of wind, water and ice. 'There is a great challenge in the sea', wrote Henry 'Hank' Stommel in 1945, 'a powerful urge which attracts and enslaves us – and there is happiness and companionship too, but we know that the sea is one domain which we must never dare hope to conquer'.[6]

This book explores the history of the circumpolar ocean at the southernmost end of the world, taking us on a voyage across its wild and contested waters and into its mysterious depths. On one level it tells a global story about changing perceptions of oceans and their vulnerability to human actions. On another it is a local story, revealing how the Southern Ocean has defined and sustained places and people in the Southern Hemisphere. It explores different knowledge systems that have shaped scientific and cultural understandings of the Southern Ocean, and considers some of the imperatives for developing a deeper knowledge of this remote and stormy realm. It is not a comprehensive history of human engagement with the ocean. Rather, it interweaves environmental and human stories to trace the history of the ocean from deep geological time to the present day. We sail across its stormy surface, navigating its winds and reading its waters. We encounter the Antarctic Circumpolar Current as it journeys around the globe, tracing the movements of ideas and people and species caught up in its flow. Along the way we delve into the ocean's depths, map its geographies, hear ancestral stories of its peoples, explore ideas about its marine environment and view some of the revolutionary discoveries that have emerged from its scientific voyages. Ultimately, this book seeks to create a broader awareness and appreciation of the history and environment of the little-known circumpolar ocean of the Southern Hemisphere, and of its emerging importance as a barometer of planetary climate change.

I

OCEAN

Even now my heart
Journeys beyond its confines, and my thoughts
Over the sea, across the whale's domain,
Travel afar the regions of the earth

'The seafarer', eleventh century[1]

Latitude 46° 35′ South, Longitude 168° 19′: Bluff, New Zealand (2 October 2016)

The morning air is quiet and cool as we reach the lookout above the tiny seaport of Bluff, on the southernmost point of the South Island of New Zealand. A heavy fog blankets land and sea, smoothing out the ragged coast and making a mockery of our polar-rated jackets. A yellow sign bravely points the way to the South Pole, a mere 5000 kilometres across the Southern Ocean. A breeze rises, ruffling the sea as if to remind us that this is the domain of the Roaring Forties. Slowly, imperceptibly, Stewart Island emerges from the white folds of moist air. On its far side lies South West Cape, one of the five great sentinel capes of the Southern Ocean.[2]

The American nature writer Henry David Thoreau had a deep love of the ocean. Having spent his childhood on the coast of New England, he liked to ponder the ocean's infinite presence and its power to transcend human boundaries of time and space. He was struck by its fluid and interconnected nature, and the way that boundaries were blurred between ocean and land, sea and atmosphere. Thoreau made three visits to Cape Cod between 1849 and 1855 to 'get a better view' of the ocean, as he put it in a book published in 1865. Inspired by stories from the ancients, he imagined the ocean as the 'laboratory of continents' – the origin of all things – in the manner of the Greek legend that the land rose out of the ocean in a frenzy of creative chaos. He also drew on recent scientific insights, such as those contained in the German-Swiss geologist and naturalist PJ Edouard Desor's 1849 treatise 'The ocean and its meaning in nature' and Charles Darwin's *On the Origin of Species*, of 1859, concluding that the ocean was inhabited by such an abundance of life that, in comparison, 'our most thickly inhabited forests appear[ed] almost as deserts'. For all its fertility, however, the ocean remained an alien place where humans were out of their element. For Thoreau, the deep ocean was 'wilderness reaching round the globe, wilder than a Bengal jungle, and fuller of monsters'. He continued, 'We do not associate the idea of antiquity with the ocean, nor wonder how it looked a thousand years ago, as we do of the land, for it was equally wild and unfathomable always.'[3]

Ocean is our planet's most prominent feature. It covers seven-tenths of Earth's surface and contains about 97 per cent of its water.[4] Modern maps generally show five major oceans: the Atlantic, Pacific, Indian, Arctic and Southern (or Antarctic). In reality they form parts of the vast interconnected body of salt water covering the planet's oceanic crust. The word 'ocean' entered the English language via the Latin *ōceanus* from the Greek *ōkeanós*, meaning 'great stream encircling the earth's disc'.[5] Okeanós was the first-born son of Uranus (heaven) and Gaea (Earth) and the father of 3000 stream spirits and 3000 ocean nymphs. The Greek author Homer described him as the begetter of the gods and of all things.[6] Here lie the seeds of Western belief in the 'eternal sea', an immortal realm existing before human life and continuing to exist thereafter. The classical Greek philosopher Plato theorised that the ocean's waters passed through the centre of Earth, creating a deep reservoir that swelled with the tide and distributed water to the rivers and seas at the surface. The ocean was the 'supreme enigma', an idea that shaped Western writings for centuries.[7]

With the rise of Christianity came biblical Creation narratives about the oceanic environment. According to the Old Testament book of Genesis, the ocean was created when God opened up the abyss of waters after the great flood, and it represented 'the remnant of that undifferentiated primordial substance on which form had to be imposed so that it might become part of Creation'.[8] The ocean could not be tamed by humans in the way that they had been able to transform wild forests into agricultural landscapes. In Western societies influenced by Graeco-Roman cultural traditions, the ocean was an unfathomable mystery, an alien environment inhabited by creatures beyond the human domain.[9]

New interpretations of the Creation story began to emerge in the eighteenth century to explain the natural history of Earth. In *The Sacred Theory of the Earth*, first published in 1681, the Anglican clergyman Thomas Burnet proposed that the containment of the great flood had left behind the ragged coastlines and reefs and rocks that characterised ocean environments, while debris from the flood had settled at the bottom of the great abyss, making the ocean bottom 'so deep, and hollow, and vast; so broken and confus'd, so every way deform'd and monstrous'.[10] It is not difficult to understand that, for sailors on long-distance voyages experiencing wild tempests and the ever-present threat of shipwreck far from home, the ocean was a place to be feared. Indeed, it was often depicted in Christian cosmology as a maelstrom, a damned world haunted by drowned souls and the domain of monstrous creatures.

The story of the Southern Ocean begins with the long, slow dance of Earth's crust as it shifted and compressed and fractured over the course of 3.5 billion years. By about 300 million years ago the crust had assembled into a single landmass, Pangaea, surrounded by a single ocean, Panthalassa. When Pangaea began to split apart about 250 million years ago, two supercontinents emerged: Laurasia, which drifted northward carrying present-day Europe, Asia and North America; and Gondwana which moved southward, carrying components of the future southern continents including Australia, New Zealand, Antarctica, South America, Africa, Arabia, Madagascar and the Indian subcontinent. Gondwana too began to break apart, from around 165 million years ago. Finally, around 40 to 20 million years ago a rift opened between the southernmost landmasses, Australia and Antarctica. Ocean water flooded in from the west, marking the end of a continental bond that had lasted for a billion years and heralding the birth of the Southern Ocean. Australia slowly drifted northward, with its raft of Gondwanan species, while a deep water channel formed around Antarctica beginning with the opening of Drake Passage, the 800-kilometre stretch of ocean between Cape Horn and the Antarctic Peninsula named after the English sea captain and privateer Francis Drake in 1578. A circumpolar ocean current closed in on the now-isolated Antarctic continent. Cold polar air and ice claimed this southern land, and all but the most resilient of its once-prolific Gondwanan life forms were extinguished. Sea levels began to rise around 20 000 years ago and, over several thousand years, finally drowned the isthmus connecting Tasmania to the Australian mainland, forming the shallow sea of Bass Strait.[11]

In 1487–8 Bartolomeu Dias, a Portuguese navigator and explorer, led an expedition around the southern tip of Africa in a quest to find the southern limit of the continent and expand the reach of the Portuguese Empire by finding a navigable sea route from Europe to South Asia via the Atlantic Ocean. It did not begin well. After landing at the coastal town of Angra Pequena (now Luderitz) in Namibia, Dias's ships became engulfed in a violent storm and sailed away from the coast, out of sight of land. Days later they found shelter in a bay, having rounded the cape without seeing it. Dias had accidentally found his way into the Indian Ocean, and it held the promise of a new world far beyond the familiar oceans and seas of Europe. In fact, Arabic and Portuguese traders had mastered the traditional navigational method of coasting to explore the coastlines of Africa and South America and had been trading along this route for centuries, tapping the power of the summer monsoons to propel them eastward around the cape and into the Indian Ocean then catching the East African coastal current northward into the Mozambique Channel. Based on first impressions, Dias named the heel of land along the southernmost coastline of Africa, at latitude 34° South, Cabo Tormentoso (Cape of Storms), but after encountering strong currents he decided to turn back at Great Fish River. This time he sighted the cape. Apparently pleased with this navigational achievement, an omen that India could be reached by sea, John II of Portugal reputedly renamed it Cabo da boa Esperanza (Cape of Good Hope), and this name resonated for centuries afterwards in the journals and letters of European circumpolar voyagers who rounded the cape en route to southern colonies and the Antarctic region.[12]

The southern tip of South America proved to be an even greater navigational challenge. Stretching into the high southern latitudes, this region lies in the path of powerful westerly winds and currents. When the Portuguese explorer Ferdinand Magellan found a navigable passage through the continent between the Atlantic and Pacific oceans during his voyage around the world in 1519–22, he became the first European to reach the archipelago of Tierra del Fuego (Land of Fire), the southernmost inhabited land in the world. It was another century before two Dutch explorers, Jacob Le Maire and Willem Schouten, rounded the rocky headland on Hornos Island in Tierra del Fuego, which they named Kaap Hoorn (Cape Horn), successfully negotiating that infamously stormy passage in 1616.

Writing in 1697 after a 12-year voyage around the world, the explorer William Dampier advised English trading ships bound for the Philippines to set out at the end of August and sail westward into the Pacific Ocean, passing around Tierra del Fuego. Once past the Horn they could then take advantage of the 'constant brisk' easterly trade winds. Not only would this route reduce their voyage by several weeks, to six or seven months at most. It would also mean they could avoid passing the Dutch settlements around Batavia (present-day Jakarta) in the Dutch East Indies (now the Republic of Indonesia).[13]

Those who successfully navigated Cape Horn were met with the unfamiliar winds and unpredictable currents of the Southern Ocean. The admiral of the British fleet George Anson found himself in the ocean's thrall when he led a squadron of six vessels around the cape in 1741 in search of Spanish treasure ships in the Pacific Ocean. Anson ordered his squadron to take shelter from the 'continual terror' of the Southern Ocean by keeping close to the coast of Tierra del Fuego, but the vessels found themselves caught in the maelstrom of the Drake Passage as winter approached. Anson later wrote, 'The violence of the current, which had set us with so much precipitation to the eastward, together with the force and constancy of the westerly winds, soon taught us to consider the doubling of Cape Horn as an enterprize, that might prove too mighty for our efforts.'[14] In the 'inhospitable latitudes' of wind, hail, rain and massive waves, seamen were washed away, rigging and masts destroyed. Anson lost sight of two of his ships in thick fog and never saw them again. These navigational difficulties were compounded by limited geographical knowledge and the ever-present dangers of scurvy, mutiny, piracy and shipwreck.[15]

Mariners' journals and voyagers' letters and diaries offered vivid personal testimonies of such journeys. For early voyagers the unfamiliar ocean, as the French historian Alain Corbin wrote, was like 'a road without a road, on which man drifts in the hands of the gods, under the permanent threat of hostile water, which is a symbol of hatred that extinguishes the passion of love as it does fire'.[16] Such accounts immersed readers at home in the sense of adventure and excitement of voyaging in the southern seas without ever having to leave their armchairs. The editor of a collection of British sailors' experiences in the Southern Ocean published in 1827, for example, suggested that readers interested in literature, science and the arts would find the stories exhilarating.

> The vicissitudes of a life at sea are more striking, and calculated to excite a deeper interest, than any other which the circle of real life presents. The continued change of scene, and the extreme peril which every moment impends over the mariner, render his life a scene of perpetual excitement.[17]

The oceanic imagination was flourishing, nurtured by travellers' tales and accounts of disease, mutiny, shipwrecks and ships disappearing without a trace.

George Shelvocke's 'long and unfortunate voyage' from 1719 to 1722 gave rise to one such account. The former British naval officer had turned to privateering after an old shipmate appointed him as commander of a private expedition to the Great South Sea in search of Spanish treasure ships, funded by the consortium of merchants known as the Gentleman-Adventurers.

Privateering, which began in 1243 and continued until the signing of the *Declaration Respecting Maritime Law*, in 1856, was commonplace in the eighteenth century.[18] It involved privately owned vessels, operating under the authority of the state, attacking and plundering enemy vessels. Privateering expeditions enabled states to conduct covert warfare at sea without resorting to public money to equip fleets of warships. The merchant owners and privateering crews were entitled to split 90 per cent of the profits between them, while the state in question claimed the remaining 10 per cent. Privateering manuals offered advice on the best routes and strategies for taking enemy ships, especially in the Spanish-dominated seas off South America.

Shelvocke's appointment as expedition commander was short-lived, however, when reports of unruly conduct filtered back to the expedition's sponsors who promptly appointed John Clipperton as captain of the larger ship, *Success*. Shelvocke was made captain of the smaller vessel, *Speedwell*, and subject to Clipperton's command. For most of the voyage Shelvocke nursed a seething resentment towards Clipperton.[19] They set sail for the Strait of Magellan, but, six days after leaving Britain, the *Speedwell* and *Success* were separated by a storm. The choice of route into the Southern Hemisphere was generally a matter for individual captains to determine based on weather, currents and time of year. Shelvocke, who was without any of the expedition's charts to guide him, chose to navigate through the Le Maire Strait to reach the Juan Fernández Islands off the coast of Chile, where he intended to rendezvous with Clipperton. The islands form an archipelago in the Pacific Ocean and were named after the Spanish sailor who was the first to sail into the open ocean off Chile, in 1574, in order to avoid the strong northerly flowing Humboldt Current. On 11 January 1720 Shelvocke's vessel was wrecked on the largest island in the group, which he called 'desolate, uncultivated … situated … in the uttermost parts of the earth'. Most of his 71-strong crew mutinied, and Shelvocke resorted to building a new boat and finding ways to survive in the harsh environment. The 'tempestuous sea', wrote Shelvocke, had diminished them to a 'forlorn state'. Nevertheless, he eventually succumbed to the 'certain savage, irregular beauty' of the island, which was destined to be immortalised in one of the best-known sea novels of the eighteenth century by the British merchant and journalist Daniel Defoe.[20] Defoe based his *Robinson Crusoe* on the experiences of another sailor named Alexander Selkirk, who had demanded to be put ashore on Más a Tierra in the Juan Fernández group after arguing with the captain of a privateering vessel. Selkirk survived on the island for four years and four months before being rescued in 1709 by another privateer ship under the command of William Dampier, whose own account of voyaging into the Southern Ocean via the Cape of Good Hope, published in 1697, was hugely popular and ran to four editions within the first two years.[21]

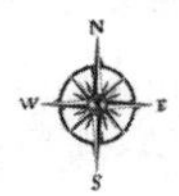

As early as the second century CE the Alexandrian astronomer Claudius Ptolemy (or Ptolemaeus) had accepted the classical theory that Earth was spherical, but he proposed that it consisted of a number of seas surrounded by land.[22] Accepting the idea that Earth was spherical, however, necessitated understanding how it maintained its equilibrium as it floated in space. Ptolemy's explanation was that a landmass existed in the southern half of the planet with sufficient weight to counterbalance that of the landmasses of Europe, Asia and Africa in the north. He laid out his theory of Earth's geographical features, including the Great Southern Land, in his eight-volume thesis, *Geography*, which provided the basis for European philosophical and geographical knowledge of Earth's physical nature for centuries. His work found fertile ground when it reached Italy from Constantinople (now Istanbul) in around 1400. The Spanish navigator Marco Polo had already travelled as far as China some 200 years earlier, bringing back accounts of new lands with incredible treasures as well as elephants and game for the taking. The account of his travels was published in German in the same year that Ptolemy's *Geography* reached European readers. While there was some confusion about the precise location of the lands that Marco Polo described, his journey became linked in the European imagination with a golden province in the southern seas.

In 1570 Abraham Ortelius produced one of the earliest modern world maps, *Theatrum Orbis Terrarum* (Theatre of the world), with *terra australis nondum cognita* (elsewhere called *terra australis incognita*, both meaning 'the unknown south land') anchoring Earth at its southern extremity and occupying the region now known as the Southern Ocean. Cartographers' maps and charts of the newly discovered South Seas followed suit, portraying a huge landmass stretching around the globe at the South Pole in accordance with Ptolemy's revered theory. Although such a land had not actually been sighted by European navigators, it filled what would otherwise have been an empty ocean.[23]

Scottish-born Alexander Dalrymple, the British Admiralty's first hydrographer, was a passionate advocate of Ptolemy's theory. Dalrymple believed that the southern land extended northwest from Tierra del Fuego and was larger than Asia.[24] He published a hypothesis about the Great Southern Land in 1767 drawing on philosophical and geographical theories as well as reports and charts from European navigators who had found their way into the southern Atlantic and Pacific oceans with the continent as their goal. These included the Dutch naval commander and cartographer Dirck Gerritsz who, in 1599 aboard the *Annunciation*, was driven far southward of Cape Horn, possibly as far as latitude 64° South. According to the Dutch explorer Le Maire, Gerritsz saw 'very high mountainous land, full of snow, as the land of Norway, very white covered', and his supporters argued that this was the first sighting of Antarctica.[25] There were two sightings, at latitude 50° South and 41° South respectively, by crew members aboard the *Orange* in 1624.[26] There were many more assumed sightings over the next century. During a French naval expedition, on 1 January 1739 Jean-Baptiste Charles Bouvet de Lozier saw what he thought was a promontory of the southern landmass and named it Cap de la Circoncision (Cape Circumcision, now Bouvet Island) after the feast celebrated on that day, only to spend 12 frustrating days trying to bring his ship to shore in dense fog. With his crew starving and succumbing to scurvy, he finally abandoned any hope of landing or going further southward and instead steered eastward along the edge of vast fields of floating ice, convinced that the unknown continent must lie somewhere to the south because of the large populations of penguins and seals in the area. Upon returning to France six months later, Bouvet informed his superiors that the southern continent was much further away than anticipated, and it was therefore unsuited for use as a resupply post for French vessels sailing to South Asia. For Dalrymple, these discoveries simply added weight to his firm belief in the existence of the unknown southern land.[27]

By the late eighteenth century the southern land had become an undisputed fact in European cartography, as James Cook's biographer the New Zealand historian John Cawte Beaglehole observed.

> Symmetry demanded it, the balance of the earth demanded it – for in the absence of this tremendous mass of land, what, asked [Gerardus] Mercator, was there to prevent the world from toppling over to destruction amidst the stars? The great southern continent was to most thinkers of the time more than mere knowledge founded on discovery and experience – it was a feeling, a tradition, a logical and now even a theological necessity, a compelling and inescapable mathematical certitude. Its discovery must come.[28]

Beyond the philosophical attributes, the promise of a new and unclaimed southern land was irresistible to the imperial powers of Europe. In 1744, the British writer, scientist and Anglican priest John Harris, in the second edition of his *Complete Collection of Voyages and Travels*, addressed 'the Merchants of Great-Britain' and claimed that the discovery of *terra australis incognita* would yield many benefits, including increased naval power and mercantile trade.[29] The Seven Years War between France and Britain in 1756–63 left a lingering British–French rivalry over new lands to colonise, and enthusiasm grew for voyages to claim the fabled southern land. There were some in circles of power who even promoted the idea that the Great Southern Land was a kind of El Dorado – a place promising fabulous wealth and opportunity – such was the competition to discover and claim new territories and control any new shipping routes that might bestow commercial advantage.

At Dalrymple's behest, the British Admiralty appointed James Cook to command a voyage to the southern seas. The southernmost end of Earth was about to shift from a distant and imagined realm to tumultuous reality. In May 1768 Cook sailed from Britain aboard the barque *Endeavour* with instructions to observe the transit of Venus in Tahiti. For Cook it was the first of three voyages into the high southern latitudes. Cook also had in his possession a second set of orders, handed to him under seal. His unpublicised mission was to ascertain whether a continent existed in the high southern latitudes and, if so, to claim it in the name of the king of Great Britain.[30]

> You are to proceed southward in order to make discovery of the continent above mentioned until you arrive at the

> latitude of 40°, unless you sooner fall in with it. But, not having discovered it or any evident signs of it in that run, you are to proceed in search of it westward … until you discover it, or fall in with the eastern side of the land discovered by Tasman and now called New Zealand.[31]

The *Endeavour* left Plymouth in August 1768. Cook's first port of call was Brazil, then he sailed south to round Cape Horn and enter the Pacific Ocean, reaching Tahiti in April 1769 from where he observed the transit of Venus. The next leg took the ship as far as latitude 40° South, but, with no landmass in sight, Cook continued on to New Zealand, circumnavigating the coastline and establishing that it was made up of two major islands. Heading westward, he sighted the east coast of New Holland (Australia) in April 1770 and charted it as he steered northward to Cape York and on to Batavia (Jakarta) before finally reaching Britain in July 1771 after a voyage of almost three years.

In a postscript to his journal, Cook outlined his plans for a second voyage to finally settle the question of the Great Southern Land. He was nothing if not tenacious. The Admiralty agreed and Cook, having been promoted from lieutenant to captain, departed Plymouth in 1772 with the vessels *Resolution* and *Adventure* under his command. This time he avoided the notoriously stormy sea around Cape Horn in favour of the easterly route around Africa, taking advantage of the strong westerly winds south of the Cape of Good Hope. His instructions were to sail to the cape and take on supplies before proceeding southward to locate Cape Circumcision, which Bouvet had recorded at latitude 54° South and longitude 35° East, and to determine whether it was part of the Great Southern Land. As Cook noted,

> If it proved to be the former, I was to employ myself diligently in exploring as great an extent of it as I could, and to make such notations thereon, and observations of every kind, as might be useful either to navigation or commerce, or tend to the promotion of natural knowledge. I was also directed to observe the genius, temper, disposition, and number of the inhabitants, if there were any, and endeavour, by all proper means, to cultivate a friendship and alliance with them; making them presents of such things as they might value; inviting them to traffic, and shewing them every kind of civility and regard.

If Cape Circumcision proved to be part of an island, however, Cook was to continue southward then eastward in search of the continent, circumnavigating the globe as near as possible to the South Pole and charting any islands he found in 'that unexplored part of the Southern Hemisphere'.[32]

The concept of fixing a position on Earth's surface using imaginary lines parallel to the equator was well known by this time. Latitude could be measured with a reasonable level of accuracy using a compass and cross-staff, a simple device used to measure the elevation angle of the noontime sun above the horizon.[33] Determining longitude at sea was far more difficult, however, since it required an accurate means of measuring time on a moving vessel. Maritime nations were keen to solve the problem, since any miscalculation of a ship's position at sea could, and often did, spell disaster.[34] They began offering financial incentives from the early sixteenth century. In 1761 an amateur British clockmaker, John Harrison, made a breakthrough with his portable marine chronometer, a spring-driven clock known as H4. It could measure the degrees of longitude traversed during a sea journey by comparing local time at high noon to the absolute time set on the chronometer at the start of the voyage. Cook took an accurate copy of Harrison's invention, made by Larcum Kendall and called K1, together with three devices made by John Arnold for testing during the voyage. The British mathematician and astronomer William Wales, who was sent on Cook's voyage by the Board of Longitude to perform astronomical observations and report to the Royal Observatory at Greenwich, was also responsible for testing the accuracy of the experimental marine timekeepers under a variety of conditions on the open ocean. Arnold's timepieces eventually succumbed to the freezing conditions, but K1 survived. According to Cook, 'Mr Kendall's Watch exceeded the expectations of its most zealous advocate and by being now and then corrected by lunar observations has been our faithful guide through all vicissitudes of climates.'[35]

The problem of longitude may have been overcome, but the Southern Ocean's notorious weather still prevailed. By December 1772 the *Resolution* was floundering in tempestuous conditions and being driven far eastward of Cook's intended course. He dismissed any hope of locating Cape Circumcision. Instead, he found himself dealing with a dramatic drop in air temperature at latitude 48° South and sighted his first iceberg at 51° South. As was customary, Cook kept a daily journal to record navigational information as well as details of the voyage, from the disposition of the crew to particular events or sightings deemed of interest to his sponsors. His observations about the physical nature of the Southern Ocean were of particular interest to the British Admiralty, especially given the fragmentary information available to navigators about sailing conditions in the high southern latitudes. One of Cook's more frequent notations concerned the size of waves in the Southern

Ocean, with swells ranging from 'very high' and 'great' to 'vast' and 'prodigious'.[36]

With the benefit of satellite technology, we now know that Southern Ocean waves regularly reach four to five metres in height and can swell to twenty metres during low-pressure systems. During a scientific research voyage in 2002, for example, the master of Australia's Antarctic research and resupply vessel *Aurora Australis* recorded swells averaging twelve metres in height during a storm, with individual waves reaching up to sixteen metres. In 2017 a newly installed buoy moored in the ocean off the subantarctic Campbell Island, 600 kilometres south of New Zealand's South Island, recorded a wave 19.4 metres high. It was one of the largest single waves ever recorded in the Southern Hemisphere.[37]

For Cook, however, the giant swells were only the beginning of the Southern Ocean's challenges. At latitude 51° South the vessels caught sight of their first icebergs and were soon surrounded by an 'immense field of Ice'. They turned north from 55° South then circled south again, forever surrounded by floating ice which Cook surmised was what Bouvet had mistakenly thought was an island. At 61° South Cook ordered three boats into the ice field to collect loose pieces that could be melted down for fresh water, then, at 66° South, the ships crossed the Antarctic Circle, the first European vessels to do so. Enveloped in a thick blanket of fog, the captains found it exceedingly difficult to avoid colliding with the ice islands. Somewhere in the vicinity of the Kerguelen Islands the *Adventure*, under the command of Tobias Furneaux, lost contact with the *Resolution* and was forced to sail on alone to Tasmania before heading to the prearranged meeting point of Queen Charlotte Sound, at the northern end of New Zealand's South Island. Cook, meanwhile, sailed southeast tracing along the edge of the ice field, before also making for Queen Charlotte Sound where the two ships were reunited in early 1773. After several months exploring the southern Pacific Ocean the two ships were again separated. This time Furneaux headed north, returning to Britain a year before Cook who steered the *Resolution* southward to resume his search for the southern continent. He crossed the Antarctic Circle for the second time before turning north to avoid the freezing conditions. Not one to be deterred, he then made another great southerly loop, crossing the Antarctic Circle for the third time on 26 January 1774.

On board the *Resolution* was a Prussian-born naturalist, Johann Reinhold Forster, whose knowledge of languages and writings on natural history had brought him recognition amongst fellow scientists, including Joseph Banks and Daniel Solander, who had supported his election to the Royal Society in 1772. For Cook's first voyage the British Admiralty had chosen Banks, 'a Gentleman of large fortune … well versed in natural history'. Banks had completed the voyage assisted by a staff of eight, including two naturalists and two artists whose role was to observe and record the plants, animals, landscapes and people encountered during the journey. Banks expected to accompany Cook on the second expedition too. Forster had written to Banks expressing a desire to accompany him on the return journey to study nature, but Banks, aware of Forster's reputation for being quarrelsome, had been unwilling to have him as a travelling companion. When the Admiralty had then refused to meet Banks's demands for staff and accommodation for the second voyage, he and Solander withdrew. Forster's application subsequently found support from the first lord of the Admiralty, John Montagu, 4th Earl of Sandwich.

Aboard the *Resolution*, and armed with a library of eighteenth-century journals of voyages and natural history, the 42-year-old pastor-turned-naturalist immersed himself in his work assisted by his son, Georg, a talented artist. Together with William Wales and the British painter William Hodges, the Forsters represented the 'experimental gentlemen', as the seamen called them, on the voyage.[38] Forster was familiar with the methods of natural history developed by Carl Linnaeus, the Swedish naturalist who had created a methodical system of identifying, naming and classifying organisms and one of the first scientists to describe the relationship between living things and their environments. In fact, Forster considered himself to be 'a kind of Linnaean being' and was more concerned with questions of natural history for the sake of advancing scientific knowledge rather than as a source of nationalistic pride in discovering new lands or species.[39] Linnaeus himself had given Forster his blessing, telling him that he demonstrated a spirit rivalling that of the heroes of war, and that his participation on the voyage would 'turn the eyes and minds of all botanists in your direction'.[40]

While Cook's two vessels had been anchored off the Cape of Good Hope for three weeks to take on fresh supplies, Forster recruited the assistance of Anders Sparrman, a Swedish botanist who had studied under Linnaeus. Beyond the cape the *Resolution* soon entered a region of treacherous gales, thick fog and floating ice. Forster proved to be a diligent observer but a fractious passenger. He complained bitterly about his cramped accommodation, which flooded in the mountainous seas, and expressed his dismay that men of science such as himself, from 'the most enlightened nation in the world', should receive such treatment. He was miserable with rheumatic pain and blamed Cook for prolonging his suffering in the gruelling conditions.

> Our whole course, from the Cape of Good Hope to New Zealand, was a series of hardships, which had never been experienced before … We had the perpetual severities of a rigorous climate to cope with; our seamen and officers were exposed to rain, sleet, hail, and snow; our rigging was constantly encrusted with ice, which cut the hands of those who were obliged to touch it; our provision of fresh water was to be collected in lumps of ice floating on the sea, where the cold, and the sharp saline element alternately numbed, and scarified the sailors' limbs; we were perpetually exposed to the danger of running against huge masses of ice, which filled the immense Southern ocean.[41]

While admiring Cook's skill as a navigator and seaman, Forster was disparaging about the commander's failure to appreciate the importance of the study of nature, and criticised his fellow passengers and crew for actively hindering his attempts to gather information and collect specimens. 'The world will, however, derive one advantage from this proceeding', wrote his son, Georg. 'We shall have little to offer, but what we have seen with our own eyes, and for the truth and precision of which we can be answerable.'[42]

To Forster's credit, he persevered with his work throughout the four-month voyage in the high southern latitudes, recording his observations of the ocean, theorising about its origins and discussing whether the phenomena he witnessed might offer clues as to the existence of a southern continent. He noted changes in the depth, colour, saltiness and temperature of the ocean water and contemplated at length the character and significance of the large islands of ice they encountered, even in mid-summer. On 26 December 1773, he counted 186 masses of ice on the horizon.

The question of how ice formed in the Southern Ocean was central to the debate over whether a Great Southern Land existed, and Forster was not afraid to take on even the most respected philosophers on this issue. An outspoken advocate of personal examination of the natural world, he was dismissive of those who constructed theories about new lands without having ever visited them. Indeed, he saw it as his mission to use his first-hand observations as a 'learned empirical traveler' to challenge the views held by some of his contemporaries about the nature of the high southern latitudes. Forster gave vent to his dislike of 'this ill founded opinion':

> Philosophers not content with quoting authorities & Discoverers of the Continent, and asserting that it was impossible to *discover* it hereafter, because it had not been seen before, tortured the imagination to invent mechanical & mathematical reasons to demonstrate the absolute necessity of Land in the Southern Hemisphere, and declared the World could not perform its revolutions without that due proportion of Earth to counterbalance the Solid Weight of the Northern half ... A mountain of Ice seen in the meridian of the Cape of Good Hope, and taken for a Promontery [*sic*], came very opportunely to the assistance of these presuming Geographers, who now extended their Continent all along the Skirts of the Southern *Atlantick* Ocean.[43]

If the 'stupendous quantities' of ice encountered in the high southern latitudes surpassed his expectations, the ice's formation puzzled him. In his published account of the voyage he discussed the problem at length, drawing on his own experiences, extensive reading and conversations with sailors on the *Resolution* who had sailed in the North Atlantic.

The French naturalist Georges-Louis Leclerc, Count de Buffon, 'the most ingenious and the most elegant writer of Natural History' according to Forster, had noted that the sea at latitude 79° North did not freeze despite its proximity to the North Pole. Buffon had argued that, since salt water did not freeze, the great masses of ice observed in the Southern Ocean could not have formed in the sea. They must therefore have been carried down to the sea by rivers or, if the South Pole was so cold that it congealed the surface of the sea, then land must exist for the ice to 'fix on' before breaking away to float in the ocean. Forster dismissed Buffon's argument that the ocean did not freeze in high latitudes, presenting numerous examples to the contrary and speculating that the ocean beyond 71° South was likely to be sufficiently cold for an extended period to congeal salt water and form extensive masses of ice without requiring land. Indeed, Forster concluded, the absence of land in the high latitudes of the Southern Hemisphere made the Antarctic region colder than the Arctic region in corresponding latitudes. In a final flourish, Forster stated that, unlike the northern seas, there was no drifting wood to be found 'in all the Southern seas', proving – if any further proof was needed – that no 'Austral' land existed that was capable of supporting vegetation.

Forster's observations of the Southern Ocean also served to validate another of his theories. Like many observers of the natural world in the late seventeenth and eighteenth centuries, during the so-called Age of Enlightenment, he believed that Earth's physical variations represented a hierarchy: as one travelled south from the fertile tropics, the world became progressively more desolate and sterile. Drawing on Linnaeus's system of classifying animals and plants, Forster described what he saw as the progressively debased nature of Earth's southern extremities, manifested in the gloomy, sparsely vegetated coastlines and the torpid and slow-moving seals and penguins languishing on the ice islands of the Southern Ocean. He applied his theory with similar zeal to the indigenous peoples of the region. Affording the Tahitians of the Society Islands in the South Pacific Ocean the 'highest rank', he classified the 'cannibals of New Zealand' as above the natives of Van Diemen's Land (Tasmania), who, in turn, he considered superior to 'the most unhappy wretches of Tierra del Fuego'.[44]

On 30 November 1773, as Cook steered the *Resolution* across the Antarctic Circle for the second time, he reached 71° 10′ South and passed close to the Antarctic continent without sighting it. Finding his ship's path blocked by a vast field of solid ice, he wrote,

> I will not say it was impossible anywhere to get farther to the south; but the attempting it would have been a dangerous and rash enterprise, and what, I believe, no man in my situation would have thought of. It was, indeed, my opinion, as well as the opinion of most on board, that this ice extended quite to the pole, or perhaps joined on some land, to which it had been fixed from the earliest time; and that it is here, that is to the south of this parallel, where all the ice we find scattered up and down to the north, is first formed, and afterwards broken off by gales of wind, or other

> causes, and brought to the north by the currents, which we always found to set in that direction in the high latitudes.[45]

Forster wholeheartedly agreed. The voyage had demonstrated conclusively that the Great Southern Land was no El Dorado, since no land apart from 'inconsiderable fragments' could be claimed for human occupation beyond 60° South. Forster concluded that even 'the learned and ingenious Charles de Brosses', a French magistrate and scholar, who had argued that such a continent must exist in order to 'counterpoise' the weight of the lands in the Northern Hemisphere, was wrong. The circumnavigation had proven beyond challenge that there was no land sufficient to balance the northern lands, although Forster cautiously added that 'nature [had] provided against this defect' in some way not yet known. He was no doubt immensely gratified to be able to contradict the promoters of the Great Southern Land. Nevertheless, frustrated by the long months spent traversing the high southern latitudes in search of the great southern continent in conditions that were hardly conducive to collecting botanical specimens, Forster wrote in his journal, 'Instead of meeting with any object worthy of our attention after having circumnavigated very near half the globe, we saw nothing, but water, Ice & Sky.'[46]

Upon his return to Britain Forster became engaged in a bitter dispute with the Admiralty over the right to publish his journal. Such long-distance voyages were often surrounded with intrigue and naval authorities, mindful of the rivalry between maritime nations and the need to control the information that expeditions might yield, generally impounded voyage logs and journals for closer scrutiny before publication. Predictably, the Admiralty refused to allow Forster to publish his own account of the voyage until the official volumes had been produced. Forster sidestepped the ban by authorising his son, Georg, to publish the journals on his behalf, then he left Britain for good. For all his diligence as a naturalist, his reputation as a temperamental and argumentative man overshadowed his achievements during the voyage. He spent the rest of his life based at the University of Halle in Germany, translating, editing and reviewing accounts of other voyages – including those of Arthur Phillip, John Hunter and John White – until his death in 1798. Meanwhile, Cook's official account was enthusiastically received.

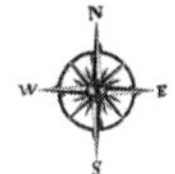

In 1794 John Rivett, the master of an academy at Wymondham, published a book for young men studying the science of geography and astronomy. He included in it 'A chronological table of remarkable events, discoveries and inventions from the Creation to the year 1794', the year in which a new series of British voyages of exploration began. The 'much lamented Captain Cook', he wrote, had traversed the Southern Ocean as far as latitude 70° South. 'This ocean is very much agitated by violent storms, and dreadful tempests; and like the northern one, contains immense islands and mountains of floating ice, that threaten destruction to all ships that approach them.' Rivett was intrigued by the physical origins of the oceans, speculating that immense volumes of water had surged from south to north, breaching areas of weakness in the coasts along the way to create new straits and gulfs, and perhaps even carving out the entire Indian Ocean between Africa and Australia. Rivett also proposed that the continents and islands had been shaped by powerful earthquakes, subterranean volcanoes and changing sea levels. 'It is certain', he wrote, 'that the globe has suffered great vicissitudes since the deluge ... so that the present inhabitants of the globe may be said to plough those lands over which ships formerly sailed, and to sail over lands which were formerly cultivated.' His words seem remarkably prescient; the modern theory of plate tectonics suggests that most of Earth's tectonic, seismic and volcanic activity occurs at the boundaries of rigid plates that are in constant motion.[47]

Cook's endeavours to find the missing continent had taken the *Resolution* closer to the South Pole than any other European vessel. He had put to rest the enduring legend of a southern land that counterbalanced the continents in the Northern Hemisphere. As Beaglehole put it, the unknown southern land 'was an illusion raised by abstract thought, buttressed by fragments of discovery that seemed to fit into a likely pattern, demolished by experienced fact'.[48] Cook had also surveyed and claimed for the British Crown the two main islands of New Zealand and the eastern coast of Australia, and charted a number of newly discovered islands in the Pacific Ocean. During his journeys in the high southern latitudes, he had noted the uniformity of the winds and currents and marine life in what he called the Southern Ocean.[49] His voyages also yielded valuable new insights into the ocean's natural phenomena and established the practice of scientists accompanying naval voyages of exploration.

Under Cook's command, the ambitions of science and empire seamlessly converged.[50] Maritime exploration and science had formed a partnership that became the hallmark of future voyages into the Southern Ocean. Perhaps Cook's most significant legacy, however, was to cast an experienced mariner's eye on the Southern

Ocean and render it physically and geographically knowable. As he wrote in his journal,

> The prevailing winds and currents in each part of the ocean are well known to us: the exact distance and bearing from one point to another are laid down in the chart; steam bridges over calm areas, and in many cases conducts us on our entire journey at a speed but little inferior to that of land travelling by railroad; modern science preserves fresh and palatable food for an indefinite period; and, in a word, all the difficulties and most of the dangers of long voyages have disappeared.[51]

Nearly 200 years later the acclaimed American marine biologist Rachel Carson hailed what she saw as Cook's crowning achievement: becoming the first European to document the existence of 'a stormy ocean running completely around the earth south of Africa, Australia, and South America'.[52] For the empires of Europe, the challenge now was to find a way to conquer it.

2

WIND

If you could master those winds, you could dominate the world.

Tom Griffiths, *Slicing the Silence*, 2007[1]

Latitude 54° 02′ South, Longitude 37° 14′ West: Prion Island, South Georgia (2 November 2017)

I used to read a book to my child called *The Wind That Blew Too Much*, and I wonder now whether the author had the westerly winds of the Southern Ocean in mind when he wrote it.[2] Early sailors called them the Roaring Forties, after the sounds they made as they howled through masts and riggings. Further south lie the Furious Fifties and, finally, the Screaming Sixties. The winds form as warm air currents rising at the equator sink, at around latitude 30° South, and travel southward over Earth's surface before rising again near the Antarctic coastline. As the temperature drops, the air is deflected southward by Earth's rotation. There is no continental landmass here to break the speed of the powerful winds, so they continue on their circumpolar journey, carving their signatures into isolated headlands and weather-beaten islands.

As I pause on the brow of a hill, I see two light-mantled sooty albatrosses wheel and soar in perfect unison. Then an apparition of an adult wandering albatross comes into view. I hold my breath. Wings locked, it circles above the golden tussac grasses and skims the currents of silver air. A small flutter, and powerful wings arch upward for a slow, poised descent to feed the waiting mouths. Fluffy chicks, half-grown, huddle close to the grass. They will soon learn to fly, testing their wings to find their inner balance. They are full of swagger despite their downy bodies. One strides over to a neighbouring giant petrel chick and picks a fight. They remind me of bored teenagers filling in time between snacks. They peck at each other's neck feathers in mock battle, then the petrel retreats to the safety of its own patch of tussac.

In 1959 Mary Gillham found herself at latitude 54° South on Macquarie Island, a tiny protrusion of oceanic crust in the midst of the Southern Ocean halfway between Australia and the Antarctic continent in what she called the 'albatross latitudes'. She was there in her role as a naturalist with the Australian National Antarctic Research Expedition (ANARE). Born in London, Gillham had served in the Women's Land Army during World War II before gaining a doctorate in island ecology. Her passion was the wildlife of isolated coastal and oceanic islands, and her research took her to some of the remotest parts of Earth. In 1958 she spent time on the islands of Bass Strait off Tasmania, studying the impact of fishing on the colonies of migratory mutton birds. In the following year she sailed to Macquarie Island on the Danish polar vessel *Thala Dan* with three other female scientists: Isobel Bennett, Susan Ingham and Hope Macpherson. In the midst of the circumpolar storm track of the high southern latitudes, she found four species of albatrosses nesting on the steep, rugged island cliffs and one of the largest congregations of seabirds anywhere in the world. Gillham was indeed a long way from home.[3]

The albatross has long been admired for its beauty and power. The wandering albatross (*Diomedea exulans*) is the largest of the dozen or so species that make up the family Diomedeidae, named by the Swedish taxonomist Carl Linnaeus after Diomedes, the Greek warrior of the Trojan War.[4] According to the myth, Diomedes angered the goddess Aphrodite by wounding her while his companions goaded her. She took her revenge by conjuring a storm that wrecked his fleet and, according to one version of the story, turned his men into large white birds and exiled Diomedes by preventing him from returning to his homeland.[5] When young adult wandering albatrosses leave their nests to take to the air, they spend up to ten years riding the circumpolar winds before beginning to revisit their native islands every two years to breed. With long, narrow wings forming a huge span, and the ability to lock those wings at elbow and shoulder, they have made this turbulent realm their own. Satellite tracking has shown that they can fly up to 15 000 kilometres without landing and are capable of reaching speeds of 80 kilometres per hour.[6]

The albatross is deeply revered in Māori culture, where its name is *toroa*. Toroa was an ancestral chief who skilfully navigated his canoe across the Pacific Ocean from the islands of Polynesia to New Zealand. He is said to have worn clusters of snow-white albatross feathers (*raukura*) hanging from his ears, giving rise to the Māori saying *me te pōhoi toroa tērā, pūaho ana* (like the intensely white down of an albatross). Feathers and bone pendants were worn by Māori elders; they were believed to confer the bird's special qualities upon the wearer. Albatross wing bones served as chisels for carving traditional tattoos, and woven mats depicted *nga roimata toroa* (the tears of the albatross), showing the bird weeping for its distant homeland and referring to its ability to expel salt from its nostrils as it drinks ocean water. During the late nineteenth century the white albatross feathers became symbolic of nonviolent resistance to European occupation of confiscated land, centred on the Māori village of Parihaka, on New Zealand's North Island.[7]

The imagination of European mariners was also captured by the albatrosses, with their magnificent pure white bodies and enormous wingspans, navigating the tumultuous Roaring Forties with ease, soaring above the ships and effortlessly descending to pluck morsels left in their wakes. Scandinavian sailors believed that albatrosses carried the restless souls of drowned sailors. Others were not averse to capturing them for food or entertainment. Nevertheless, few could fail to be impressed by their avian grace and formidable survival skills in the world's windiest latitudes. The ornithologist Robert Cushman Murphy, from the American Museum of Natural History in New York, recalled seeing his first albatross in 1912 during a voyage to the Antarctic aboard the whaling ship *Daisy*. He wrote,

> I now belong to a higher cult of mortals … In the morning sunlight, flew the long-anticipated bird, even more majestic, more supreme in its element, than my imagination had

> pictured … lying on the invisible currents of the breeze, the bird appears merely to follow its pinkish bill at random.[8]

During the voyage to Macquarie Island, Mary Gillham recalled leaning over the stern rail of *Thala Dan* watching the 'noble birds' landing in the ship's wake to feed on the squid being churned to the surface and felt an uncharacteristic urge to read *The Rime of the Ancient Mariner*, Samuel Taylor Coleridge's allegorical poem written in 1797–8. The poem recounts a story, told to a wedding guest by an old seaman, of a ship that sailed southward towards the South Pole. As the temperature dropped and the ship became surrounded by floating ice, an albatross came into view through the fog and followed the ship for many days. The crew welcomed it as a good omen, but the ancient mariner shot the bird. The ship was soon becalmed, and the ocean became thick with rotting creatures. As the sailors began to die of thirst, they blamed the mariner for their plight and hung the albatross's body around his neck. He did not die but, after seven days and nights, he began to pray and the bird fell from his neck. 'Somewhat unscientifically,' Gillham later wrote, 'but in the knowledge that wind and albatrosses are associated, the poet allowed his characters to believe the albatross responsible for the wind. Its death becalmed them. "Ah wretch," said they, "the bird to slay that made the breeze to blow"'.[9]

Scholars have long debated the poem's inspiration. Some attribute it to a ballad devised by Coleridge and William Wordsworth during a walking holiday in 1797 as they pondered a way of raising funds.[10] Wordsworth had recently read George Shelvocke's account of the killing of 'a disconsolate black Albitross [*sic*]' that had been following his ship the *Speedwell* for several days. Shelvocke described how in October 1719 he was being driven far to the south of Cape Horn by bad weather and the second captain, Simon Hatley, believing the albatross to be an ill omen, shot it in the hope that its death would bring a fair wind.[11] To Gillham's analytical mind, the account was a 'figment of the poet's imagination' although, as a story about the consequences of humans violating nature, it is also a parable for our time. Wandering albatrosses are regularly found dead, hooked on baited longlines set for catching the Patagonian and Antarctic toothfish of the Southern Ocean, and fewer birds are returning to their subantarctic islands to breed. They may soon become solely the stuff of legend.[12]

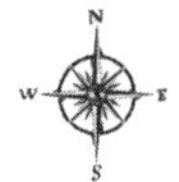

The winds of the Southern Ocean offered the ultimate testing ground for sailing prowess and physical and psychological endurance. They still do. Taking the fastest sailing route around the globe requires navigating the southernmost capes of South Africa, Australia, New Zealand and South America, capes which have long been marked on sailing charts as crucial navigational landmarks in this ocean hemisphere. Three of them – the Cape of Good Hope, Cape Leeuwin and Cape Horn – are collectively known as the 'great capes', and they have assumed almost mythical proportions in modern sailing circles. At the latitudes of these capes, sailors must be prepared for gale-force westerly winds and mountainous seas while avoiding collision with submerged rocks and drifting icebergs.

More than a century after Bartolomeu Dias's voyage in 1488, the Dutch mariner Hendrik Brouwer proposed a new sea route to Jakarta, with a view to giving vessels of the Dutch East India

Company a faster passage across the Indian Ocean. The Brouwer Route required vessels to round the Cape of Good Hope and then sail southeast, into the path of the Roaring Forties, then to navigate eastward in the Roaring Forties for about 7400 kilometres before heading northward to Indonesia. The traditional route to the Spice Islands, or the Moluccas, closely followed Africa's eastern coastline and ships were often becalmed, increasing the length of the voyage by many months. By tapping the strong westerly winds further south, vessels could sail at far greater speeds and avoid the possibility of hostile encounters with Portuguese and other trading vessels in the stretch of water between Mombasa in Kenya and India. By 1616 the Dutch East India Company had made it compulsory for their ships to use the Brouwer Route. The new route effectively halved the sailing time, but it was not without risk. At that time there was no reliable method for calculating sailing vessels' longitudinal position in the vast circumpolar Southern Ocean, and sailors had to rely on the timely sighting of land – generally Amsterdam Island or Saint Paul Island – to know when they needed to steer northward. Missing the turn could mean a hazardous encounter with the uncharted coastline of Australia. British vessels began trialling the Brouwer Route in 1621, but disaster struck in the following year when the captain of the British East India Company vessel *Tryall* miscalculated his position and overshot the turn. He attempted to change course too late, and the ship was driven onto rocks off the northwestern coast of Western Australia. The *Tryall* was the first, but by no means the last, casualty of the fast new sailing route in the high southern latitudes.[13]

The route via the Cape of Good Hope may have offered enticing prospects of a speedy eastward voyage in the embrace of the

Roaring Forties, but mariners were not convinced that it offered the safest return trip. Thirteen years after James Cook's third and final voyage into the southern seas, Britain's First Fleet arrived at Botany Bay on Australia's east coast to deposit its human cargo of convict and guard colonists. The commander of the *Sirius*, John Hunter, was ordered to immediately set sail for the Cape of Good Hope to buy provisions for the colony. The colonial governor, Arthur Phillip, was keen for Hunter to head westward from Tasmania into the teeth of the Roaring Forties, but Hunter was doubtful about the practicality of a westward return. 'From my own experience of the prevalence of strong westerly winds across that vast ocean, I am inclined to think it must be a long and tedious voyage', he wrote. The *Sirius*, showing the strain of the eight-month voyage from Britain, was ill equipped to make the journey. Hunter chose instead to sail eastward, via Cape Horn, circumnavigating the globe in order to reach South Africa. It took him more than seven months to complete the voyage back to Botany Bay.[14]

For captains of vessels sailing into the vast southern oceans relying on crude or untested navigational charts, close attention was demanded to changes in the atmosphere, in water surface conditions and in the appearance of whatever floated or flew that might give early indications of land or ice. Henry David Thoreau described how Cape Cod seamen, like the ancient Greeks, believed that seabirds could discern changes occurring in the depths of the ocean and communicate, through their cries, what was about to happen on the surface. Observation of such phenomena could make the difference between life and death.[15]

The idea that water had its own language intrigued writers like Mark Twain, who worked as a riverboat pilot on the Mississippi in his youth. He recalled learning to read the face of the water, which,

> in time, became a wonderful book … In truth, the passenger who could not read this book saw nothing but all manner of pretty pictures in it painted by the sun and shaded by the clouds, whereas to the trained eye these were not pictures at all, but the grimmest and most dead-earnest of reading-matter.[16]

More recently the British travel writer Jonathan Raban described how Polynesian mariners used traditional way-finding techniques to navigate across the Pacific Ocean without instruments, depending on their ability to read the character of the sea in its changing colours and swells and in the movement of air and birdlife. Also writing about the Pacific, the Australian historian Greg Dening found that sailors who spent their working lives at sea developed a language of the ocean that helped to render it knowable. Experienced sailors, he wrote, were 'the philosophers of "looming"', able to recognise shapes as they emerged from the haze beyond the normal limits of vision. Reflecting on the traditional knowledge of indigenous peoples in the Pacific Ocean, he noted how their languages reflected seasonal currents, star risings and settings, the colours of the water, the shapes of ocean swells and different weather patterns. Surface features wrote the language of the ocean, and it was passed on from generation to generation, binding them to the sea through traditional stories, songs and poetry.[17]

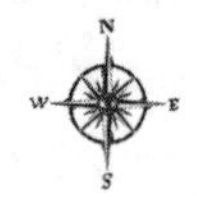

The winds of the Southern Ocean write their own legends. The southernmost capes and islands are notoriously treacherous, and the ocean's seabed is littered with many victims of its distinctive topography. One of the most enduring maritime legends about these latitudes is that of the *Flying Dutchman*. The story concerned a ghostly ship caught in a storm off the Cape of Good Hope, unable to make port and destined to sail the oceans forever. The spectre of the *Flying Dutchman* was believed to be a portent of doom to any who cast eyes upon it. The legend is thought to have originated in the seventeenth century, at a time when merchant ships were regularly navigating around the cape between Europe and Indonesia. Several versions of the legend existed. In some, the vessel itself was cursed; in others, it was the captain who was cursed, doomed to sail the oceans for eternity. While the legend's origins are unknown, it contained traces of traditional German and Dutch maritime stories such as that of Herr von Falkenberg, a sea captain who found himself condemned to sail forever on a vessel in the North Sea without a crew while he played a game of dice with the devil in return for his soul. Another story tells of a ship under the command of Hendrick Vanderdecken, who sailed from Amsterdam in 1680. Vanderdecken was said to have rashly pledged his life that he would round the Cape of Good Hope in gale-force conditions. He ignored the dangers and was condemned to repeat his perilous voyage for eternity. In yet another version the unlucky ship's master was Bernard Fokke, famous for the speediness of his voyages between the Netherlands and Jakarta. He refused to change course when caught in a wild storm off the Cape of Good Hope.

The legend of the *Flying Dutchman* first appeared in print in John MacDonald's *Travels, in Various Parts of Europe, Asia, and Africa, During a Series of Thirty Years and Upwards*, published in 1790, which inspired a flourishing trade in literary and artistic interpretations. These included Coleridge's epic poem of 1797–8,

The Rime of the Ancient Mariner, in which the mariner sights Death and Life in Death playing dice aboard a phantom ship; the Scottish writer Walter Scott's narrative poem *Rokeby*, of 1813, which tells the story of a ship unable to make port after a murder and a plague on board; the German composer Richard Wagner's 1843 opera, *Der fliegende Holländer* (*The Flying Dutchman*); and the American painter Albert Pinkham Ryder's dramatic work *Flying Dutchman*, completed by 1887.[18]

While sceptics dismissed the phantom ship legend as an atmospheric phenomenon – a mirage in which a ship on the horizon would appear to be suspended in mid-air – the tale seemed prescient to sailors embarking on voyages to far-distant oceans in pursuit of new lands and resources. In a memoir attributed to George Barrington, an Irish convict transported to Australia in the late eighteenth century, an explanation is offered of the phenomenon:

> I had often heard of the supervision of sailors respecting apparitions, but had never given much credit to the report: it seems that some years since a Dutch man of war was lost off the Cape, and every soul on board perished; her consort weathered the gale, and arrived soon after at the Cape. Having refitted, and returning to Europe they were assailed with a violent tempest nearly in the same latitude. In the night watch some of the people saw, or imagined they saw, a vessel standing for them under a press of sail, as though she would run them down; one in particular affirmed it was the ship that had foundered in the gale, and that it must certainly be her, or the apparition of her; but on its clearing up, the object (a dark thick cloud) disappeared. Nothing could do away the idea of this phenomenon on the minds

> of the sailors; and, on their relating the circumstances when they arrived in port, the story spread like wildfire, and the supposed phantom was called the Flying Dutchman. From the Dutch the English seamen got the infatuation, and there are very few Indiamen but what has some one on board who pretends to have seen the apparition.[19]

In 1880 Prince George of Wales (the future king George V) was on a three-year voyage around the world with his elder brother, Prince Albert Victor. They were both serving as midshipmen with the Royal Navy ship the *Bacchante* as part of a squadron assigned to patrol the sea lanes of the British Empire. One of the princes later claimed to have witnessed the phenomenon in the early hours of 11 July 1881 from the ship *Inconstant* as it negotiated the treacherous Bass Strait, off the southern coast of Australia.

> At 4 A.M. the *Flying Dutchman* crossed our bows. A strange red light as of a phantom ship all aglow, in the midst of which light the masts, spars, and sails of a brig 200 yards [182 metres] distant stood out in strong relief as she came up on the port bow, where also the officer of the watch from the bridge clearly saw her, as did also the quarterdeck midshipman, who was sent forward at once to the forecastle; but on arriving there no vestige nor any sign whatever of any material ship was to be seen either near or right away to the horizon, the night being clear and the sea calm. Thirteen persons altogether saw her … At 10.45 A.M. the ordinary seaman who had this morning reported the *Flying Dutchman* fell from the fore topmast crosstrees on to the topgallant forecastle and was smashed to atoms.[20]

Australia and New Zealand

Regardless of the legend's veracity, it served to remind mariners that a voyage in the Southern Ocean was not for the faint-hearted.

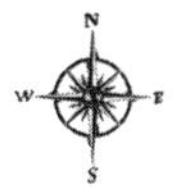

In theory, it was possible for vessels under sail to circumnavigate the globe from Cape Horn or the Cape of Good Hope without ever sighting land. In practice, many ships foundered where the massive winds and currents conspired to drive them onto rocks and shoals. The chances of a ship being wrecked while negotiating the coasts, islands, rocks and ice of the southern capes and islands were very high indeed. Added to this were incomplete charts and maps, unreliable navigational instruments, ships that were unseaworthy or simply too difficult to manoeuvre in storm conditions and the experience – or otherwise – of captain and crew.

At latitude 40° South, Bass Strait lies squarely in the path of the Roaring Forties.[21] It is a shallow basin, about 80 metres at its deepest point. It is 500 kilometres long and spans 240 kilometres between continental Australia and the island state of Tasmania. As a body of water, the strait is relatively young. Geologists estimate that it began to form as seabed movements created an east–west trough about 150 million years ago, eventually opening a passage from the west, allowing ocean water to flood in. Sea levels rose and fell many times over successive periods of glaciation, creating an intermittent land bridge across the strait that paved the way for plant and animal species to migrate. The first people are thought to have crossed the land bridge, now known as the Bassian Plain, from the north sometime before 35 500 years ago in the midst of an ice age in the Pleistocene geological epoch. Along the

Tasmanian coastline, these 'people of the Ice Age' ate a variety of foods from ocean and land and used oil from seals, penguins and mutton birds as well as land-based mammals to protect their skin from the extreme climatic conditions. The environment began to change about 12 000 years ago as sea levels rose, and the Bassian Plain disappeared beneath the waters of the Bass Strait, finally separating the Indigenous Palawa people on the island they called Trowunna (Tasmania) from other Indigenous peoples of mainland Australia.[22]

The reputation of the strait as a hazardous shipping route grew during the early nineteenth century, as increasing numbers of convict, migrant and trading vessels bound for the ports of southeastern Australia and New Zealand were obliged to tackle its winds and currents. The Roaring Forties, combined with the strait's shallow waters and scattered islands, claimed more than 700 ships over a 200-year period.[23]

The narrowest section of Bass Strait lies between Cape Otway and King Island near the western entrance, and navigating the 90-kilometre-wide passage challenged even the most experienced seamen. They called it 'threading the eye of the needle'. Many ships foundered on their way into the strait, including the convict ship *Neva* which ran onto Harbinger Reef off King Island in 1835, killing 250 people. Ten years later, in Australia's worst peace-time maritime disaster, 414 passengers and crew died after the British migrant ship *Cataraqui* attempted to sail through. Captain Christopher William Finlay had not been able to determine the ship's location in the mountainous seas, so he had calculated that the vessel was between 96 and 112 kilometres northwest of King Island by using dead reckoning, a 'best guess' navigation technique that relied on calculating the current track, groundspeed and position of a vessel based on earlier known positions. Amidst the gale-force conditions and in the darkness of night, he gave orders for the ship to be hove to, slowing its forward progress so that it gently drifted rather than having to be actively steered. But at 4.30 am the vessel crashed onto the rocks off King Island's west coast. One of the survivors later recalled how the crew helped most of the passengers onto the deck only to see them washed away in the huge seas: 'All the passengers attempted to rush on deck, and many succeeded in doing so, until the ladders were knocked away by the workings of the vessel.' By daylight the following morning around 200 people were grimly hanging on to the wreck, while others who had found sanctuary in a lifeboat were drowned when it capsized. Others died of exhaustion, still clinging to the hull as it disintegrated. According to the survivor, 'the scene of confusion and misery that ensued at this awful period, it is impossible to describe.' Of the 423 on board, only nine managed to swim or drift to King Island. They were rescued five weeks later.[24]

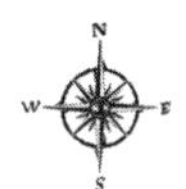

When Norwegian businessman Henrik Johan Bull ventured into the Southern Ocean aboard the whaling ship *Antarctic* in 1894–5, he described the sensation of sailing through a storm and marvelled at the hardiness of the crew as they went about their work in the wild and unpredictable conditions off Kerguelen Island.

> After a few days of this violent plunging, you get more accustomed to the complicated motions necessary to preserve a balance, and can faintly appreciate the wild beauty

> of the well-named 'roaring Forties'. The endless succession of green or gray-black mountainous billows, their breaking crests, which are blown into shreds by the squall, and carried through the rigging with a shrieking and howling as of loosened demons, the flying storm-rent clouds, and frequent mist and rain, make a picture supremely grand in its own way. With the addition of icebergs and darkness, however, I confess that a beauty of a milder type would have been sufficient for me, who had never before in person realized the astounding violence of a gale in Southern latitudes, and the rapidity with which a calm will change into a hurricane, and the latter into a calm once more.[25]

In the 1850s maritime nations began issuing to ships navigational guides containing crucial information about nautical landmarks and sailing routes compiled from information provided by local experts such as marine pilots. The British Admiralty's Sailing Directions, or Pilot Books, were the most widely used, although the first book on ocean navigation was prepared by a US naval officer for use in the northern oceans.

Matthew Fontaine Maury began his career as a midshipman in the US Navy, undertaking a circumnavigation of the globe between 1826 and 1830. After a stagecoach accident in which he broke a leg, he was unable to continue active service and was placed in charge of the Depot of Charts and Instruments, the precursor to the US Naval Observatory and Hydrographic Office. There he began the ambitious task of drawing together disparate pieces of information from captains' log books, charts and stories to produce an atlas of the world's winds and currents. When it was published in 1852, for the first time maritime nations had a picture on a world scale of tides, ocean currents and climate patterns. Maury's work inspired an international conference on maritime navigation at which nations agreed to develop, under Maury's guidance, a standardised global weather-observing system. Maury's plan was simple. Captains on all vessels at sea would be required to regularly record the weather and the oceans' physical characteristics, including sea surface temperature. In return they would receive charts of ocean currents and weather conditions relevant to their planned routes.[26]

In 1855 Maury produced a book describing the current state of knowledge about the world's oceans and climate, emphasising its practical value for faster and safer navigation. Masters of sailing vessels, he wrote, could now cross the 'pathless ocean[, and] the remote corners of the earth were brought closer together, in some instances, by many days' sail'. A knowledge of wind, weather and current made even the Roaring Forties seem less menacing:

> To appreciate the force and volume of these polar-bound winds in the Southern Hemisphere, it is necessary that one should 'run them down' in that waste of waters beyond the parallel of 40° S, where 'the winds howl and the seas roar.' The billows there lift themselves up in long ridges with deep hollows between them … The scenery among them is grand, and the Australian-bound trader, after doubling the Cape of Good Hope, finds herself followed for weeks at a time by these magnificent rolling swells, driven and lashed by the 'brave west winds' furiously.[27]

In the same year as the publication of Maury's book, the British Admiralty issued a fifth edition of its Sailing Directions for

Australia, including a chapter on navigating its extensive southern coastline. The directions drew on surveys carried out by the British navigator Matthew Flinders and the French naval officer Joseph-Antoine Raymond de Bruni D'Entrecasteaux and cartographer Louis-Claude Desaulses de Freycinet, as well as two volumes of navigation tables prepared by John Thomas Towson, a British watchmaker, in 1848 and 1849.[28] The Admiralty published Towson's tables 'to facilitate the practice of Great Circle Sailing', an idea first proposed by Sebastian Cabot in 1495, which entailed sailing the shortest distance between two points on Earth's surface during long ocean voyages. The imperative was not simply to promote safe navigation, but to save time on long-distance voyages to the southern colonies. British ship owners and merchants would reap the benefits if their ships could find a way to sail by the shortest possible route to distant ports on the other side of the globe. Towson became convinced that the fastest way to circumnavigate the globe lay in a combination of the old idea of great circle sailing and Maury's new system of navigation, a method he called 'composite sailing'. Towson predicted that his method would reduce the length of voyages to Britain's southernmost colonies by up to 20 per cent, and his timing was perfect.[29]

The discovery of gold in New South Wales in 1851 sparked a gold rush to the Australian colonies to rival that of the Californian goldfields from 1848. In setting out his arguments for composite sailing to the Liverpool Literary and Philosophical Society in 1853, Towson noted that in the previous year alone more than 200 vessels had sailed from the Port of Liverpool for the southern colonies.[30] By then the Brouwer Route was known simply as the 'clipper route' after the long, slim sailing ships bearing square-shaped rigs and three tall masts.[31] Clipper ships were the classic sailing vessels of the nineteenth century. The design is thought to have originated from the small coastal packet ships known as Baltimore clippers and evolved into that of the true clipper vessel in North American and British shipyards. With their large sail area and streamlined hulls they were built for speed, giving merchants a competitive edge on the transatlantic crossing and other lucrative trade routes. One voyage between London and Sydney in 1883 aboard the *Samuel Plimsoll*, for example, took 72 days compared to the 92 days taken by the *George Marshall* to cover the same route in 1862.[32] The clippers carried passengers as well as freight around the globe between the 1840s and 1870s. During the 1850s, meeting the increased demand for ships to carry passengers to the gold rushes, migrant clippers became a regular sight in the harbours of Britain's southern colonies. For such voyages which circumnavigated the globe via the great sailing circle, the clippers were perfectly suited to take advantage of the Roaring Forties to cover the long distances at impressive speeds. The *City of Adelaide*, purpose built in 1864 to carry passengers from Britain to the colony of South Australia, made 23 return voyages and was amongst the fastest of the clippers on the London to Adelaide route.[33]

The key to faster sailing times across the Southern Ocean lay in navigating further south, into the Furious Fifties. The higher the latitude, the shorter the route. Drawing on the experience of GB Godfrey, master of the *Constance*, Towson concluded that the fastest route for ships rounding the Cape of Good Hope was 51° South. At this latitude ships could avoid the 'rotary gales' of the Southern Hemisphere, which were known to extend as far south as latitude 48° South, and take advantage of the more constant westerly winds and rolling seas of those latitudes. 'By thus giving the Cape a wide berth', Towson wrote, 'we not only save distance, but we avoid the region of storms.'[34]

The composite route not only promised a faster voyage but also gave mariners a degree of confidence in navigating the vast stretch of ocean between South Africa and Australia. Nevertheless, such a route demanded a careful reading of wind, current, season, ice and latitude. In 1857 Towson presented another paper, to the Historic Society of Lancashire and Cheshire, this time addressing the problem of icebergs in the Southern Ocean. Robert FitzRoy, who had served as captain of the *Beagle*'s second expedition to Tierra del Fuego during Charles Darwin's voyage in 1831–6, wrote a preface to Towson's paper noting that it was by then standard practice in the 'Australian Trade' to sail in the higher latitudes.[35] As a result of Towson's work, the British Admiralty recommended that ships sailing round the Cape of Good Hope head as far south as latitude 38° South, where they could avoid the 'boisterous and stormy' weather and sudden wind changes associated with the capes and the ice associated with more southerly latitudes even in summer months.[36]

If sailing ships were beholden to the winds and the abilities of the crew to utilise them, steam-powered vessels offered far greater mastery over the ocean while also reducing the size of the crew needed for long-distance voyaging. The development of steamships during the nineteenth century was initially driven by the demand for faster, more efficient means of sea travel, particularly for commercial trade routes crossing the Atlantic Ocean. Discoveries of gold in the southern Australian colonies increased the demand for larger and faster vessels that could ply the route between Britain and Australia. By 1853 Britain was operating the steam-powered *Great Britain* between London and Australia; this was followed by the launch in 1858 of the largest ship in the world at that time, the *Great Eastern*, designed by the renowned shipbuilder Isambard Kingdom Brunel. It was equipped to carry 4000 passengers and able to sail to Australia without needing to refuel. Nevertheless, coal was an expensive commodity, and most steam-powered ships required regular stops to refill their bunkers. As a result, long-distance ocean voyaging in the Southern Hemisphere continued to rely on the power of the winds, at least until the last decades of the nineteenth century when triple-expansion steam engines, in which the steam expanded in three stages in order to extract more energy, replaced the old square riggers.[37]

Joshua Slocum knew all about the capricious nature of the great capes. A Nova Scotian adventurer, he was the first sailor to tackle the winds of the Southern Ocean single-handedly, sailing around the world in his 11-metre sloop *Spray* in the final years of the nineteenth century. As he navigated southward along the coastline of Tierra del Fuego, searching for the eastern entrance to the Strait of Magellan, he became engulfed in a storm that raged for 30 hours. 'It was now the blackest of nights all around, except away in the southwest where rose the old familiar white arch, the terror of Cape Horn, rapidly pushed up by a southwest gale.' He headed out into mountainous waves to avoid the coast, his mainsail torn to rags. On nearing what he thought was the entrance to the strait, he was startled to hear the roaring of breakers. Exhausted and overcome with seasickness, he realised too late that he had missed the entrance and instead faced a terrifying night in the 'Milky

Way' of the sea, as Charles Darwin had called this treacherous region during the voyage of the *Beagle*. Around Fury Island (Isla Furia) in Chile, northwest of Cape Horn, giant waves smashed into half-submerged rocks, creating a galaxy of white surf. Certainly, many a sailor in that part of the ocean must have imagined that he was closer to the heavens than he would have liked. On 10 June 1834 Darwin wrote, 'One sight of such a coast is enough to make a landsman dream for a week about shipwrecks, peril, and death.' To this Slocum added 'or "seaman" as well'. Throughout the night, Slocum tacked back and forth in the Milky Way, hail and sleet slashing at his exposed skin until blood trickled down his face. By daybreak he had navigated through the labyrinth of islands in Cockburn Channel and anchored in a cove, exhausted yet exhilarated, and full of praise for his yacht. 'This was the greatest sea adventure of my life. God knows how my vessel escaped.'[38] Slocum had taken almost two months to sail the 644 kilometres between the Atlantic and Pacific oceans, through the channels of Tierra del Fuego. He went on to complete his circumnavigation, covering 74 000 kilometres, in just under three years. His account of the voyage, published in 1899 and serialised in *The Century Magazine*, brought him fame and fortune. He was widely regarded as one of the last heroes of the 'age of sail' but in 1909, as he enjoyed one of his regular winter voyages to the West Indies, the sea claimed him as its own.

French-born Bernard Moitessier is also on the select list of solo sailors to circumnavigate the globe. Sailing a yacht named *Joshua* in honour of Joshua Slocum, Moitessier entered the *Sunday Times* Golden Globe Race, the first round-the-world yacht race, held in 1968. He was determined to beat the times set by ships sailing the clipper route in the nineteenth century, but alas the race was plagued by controversy. Only one contestant reached the finish line. Another disappeared at sea. A third later committed suicide. On the final leg Moitessier himself abandoned the race – and almost certain victory – when he decided to follow his own spiritual path and let the wind decide his course rather than compete for line honours.[39] He later reflected on his reason for entering the race:

> A sailor's geography is not always that of the cartographer, for whom a cape is a cape, with a latitude and longitude. For the sailor, a great cape is both a very simple and an extremely complicated whole of rocks, currents, breaking seas and huge waves, fair winds and gales, joys and fears, fatigue, dreams, painful hands, empty stomachs, wonderful moments, and suffering at times. A great cape, for us, can't be expressed in longitude and latitude alone. A great cape has a soul, with very soft, very violent shadows and colours. A soul as smooth as a child's, as hard as a criminal's. And that is why we go.[40]

Today's ocean-racing yachtsmen and women are amongst the few with any practical experience of sailing in the thrall of the Roaring Forties. While the routes for the races have varied over the years, the most prestigious are those that circumnavigate the globe, and rounding the great capes of the Southern Ocean – whether solo or otherwise – is considered to be the crowning achievement of this extreme sport.

Nearly 100 years after Joshua Slocum's voyage, an Australian woman set sail from Sydney Harbour to circumnavigate the globe on her 11-metre yacht *Blackmore's First Lady*. Kay Cottee was not competing in a race; she was raising funds for charity, and, over the course of 189 days, she became the first woman to sail solo, unassisted and non-stop around the world. Heading eastward across the Pacific in November 1987, she rounded South West Cape off New Zealand, then headed for the treacherous coastline of Tierra del Fuego. On 14 January 1988, at latitude 51° 45′ South and just 1000 kilometres from Cape Horn, she tried to slow her boat down in the face of winds of 100 kilometres per hour. The radar screen was useless because of interference from the huge waves, and she had to be constantly alert to icebergs looming into her path. This was early summer, and Cottee had all the benefits of modern communication and navigation technology, but in those latitudes exhaustion was her constant companion. Nevertheless, as her yacht plunged violently into troughs between waves 20 metres high, Cottee could still marvel at the beauty of sun shining through aquamarine water and delight in the two albatrosses that followed in her wake. Finally, on 19 January, she was simultaneously exhilarated and relieved to be safely around the great cape that had claimed the lives of so many sailors.

By April she was approaching the Cape of Good Hope, giving it a wide berth to avoid the convergence of the southerly flowing Agulhas Current and the Southern Ocean, an area of shallow water that extends 224 kilometres south of the African continent. In storm conditions, the convergence becomes a maelstrom, with waves up to 30 metres high. She later reflected that negotiating through an ocean on the lookout for icebergs was rather more preferable to sailing amongst those waves. After battling to get far enough south to avoid the shallow water over Agulhas Bank, where the Indian and Atlantic oceans converge, she received advice to turn back towards the cape because of an anticipated low system. To her dismay, the forecast proved to be inaccurate and she was forced to turn back south-southwest and into the full force of the Roaring Forties. She was exhausted and nursing a neck injury when she finally passed the cape, relieved to be plotting her course for Australia, just 7242 kilometres away, and back in the more familiar conditions of the Southern Ocean.[41]

3

COAST

Far from land, far from the trade routes,
In an unbroken dream-time
Of penguin and whale
The seas sigh to themselves
Reliving the days before the days of sail.

Derek Mahon, 'The banished gods', 1975[1]

Latitude 54° 03′ South, Longitude 37° 19′ West: Salisbury Plain, South Georgia (2 November 2017)

This is my seventh landing on the subantarctic island of South Georgia, 2000 kilometres southeast of Cape Horn. As our boat skids up onto the beach, I see snowfall tracing worry lines down the side of the ridge. The grey rock is deeply pocked by grinding ice and wind, resembling the scarred bodies of the elephant seals that line the beaches below. Tussac grass bravely climbs some of the way up then stops abruptly at the edge of a glacier. A thunderous crash of ice rocks the island as a chunk breaks away. They say some glaciers are retreating a full two metres a day. You can watch global warming in action here.

Salisbury Plain looks nothing like its British namesake. The whole scene is a study in grey and white, relieved only by splashes of gold amidst the colony of 400 000 king penguins that line the beach and low slopes, their gorgeous yellow cheeks and throats glimmering above immaculate grey-black feather capes. As I step from boat to beach, an individual separates itself from its companions and waddles up to me. I stand to attention and await inspection. Up and down, this beautiful creature slowly surveys me from head to toe. It feels like I am being greeted by a long-lost friend saying 'Is it really you?' It is hard not to anthropomorphise a king penguin at close range. Many bird species rely on geography to find each other when returning to their nests. The kings have no nests, so they must rely on sound, and the sound of 400 000 birds is surreal indeed. Each bird has two voices, a two-tone trumpeting call that, once heard, can never be forgotten. This distinctive vocal pattern helps them to identify their mate in the crowd when incubating or feeding chicks. In the noisy fraternity, communication is the key to survival.[2]

The cacophony of bellowing elephant seals, whining pups and trumpeting king penguins mingles with the sounds of the surf rolling on smooth black pebbles. It is like a wondrous choir that infiltrates the soul. The Antarctic historian Tom Griffiths wrote of the 'changeover' that occurs at the end of the Antarctic winter, when resupply ships finally breach the sea ice to reach the remote polar stations. I think that nature has its own version of changeover here, on South Georgia. The days are getting longer, and an assortment of marine mammals and birds have congregated in their

> thousands for a brief burst of frenzied mating and birthing. Antarctica may be 'a land of enveloping silence', as Griffiths put it, but the beaches of the subantarctic islands are a study in sensory overload.[3]

Tierra del Fuego, the group of islands lying at the tip of South America, was originally called Land of Smoke by Ferdinand Magellan in 1520, after the hundreds of beach fires he observed from his vessel. It was reputedly given its current name, Land of Fire, by Charles I of Spain. This most southerly inhabited land in the world lies directly in the path of the Furious Fifties, just 1000 kilometres from the Antarctic Peninsula. In December 1832 the *Beagle* reached Tierra del Fuego.[4] Robert FitzRoy, the ship's master, was 12 months into what became a five-year voyage to survey and chart the coastline of South America for the British Admiralty. This was the *Beagle*'s second voyage to the region. Aboard was Charles Darwin, who had acquired an interest in natural history as a child and was an avid collector of shells, birds' eggs, rocks and minerals. Having dismissed first medicine then geology as a career, he had been encouraged to give science another chance by his friend the reverend professor John Stevens Henslow, who recommended that he sail on this second voyage of the *Beagle*. FitzRoy agreed to take him, but as a gentleman companion rather than as a 'finished naturalist'. The Admiralty also made it clear that the study of natural history would be secondary to the main object of the voyage. Nevertheless, as Henslow put it, Darwin might indulge himself by 'collecting, observing and noting anything to be noted in Natural History'.[5]

For the young Darwin it was a once-in-a-lifetime opportunity to explore distant lands and seas. He amassed a formidable collection of animal and botanical species and recorded his impressions of the environments and peoples that he encountered along the way. He also began formulating his ideas about the mutability of species and, in particular, why some organisms survived while others became extinct. The climate, vegetation and animal life of Tierra del Fuego, Darwin observed, were quite unlike those of Patagonia to the north. Apart from whales and seals, mammal life was limited to a bat, several species of mice, tucu-tuco (a rodent), foxes, sea otters, guanaco and deer, and most of these inhabited the drier, eastern part of the country. He saw few birds in the 'gloomy woods' and no reptiles or frogs at all: 'I could not believe, that a country as large as Scotland, covered with vegetable productions, and with a variety of stations, could ever have been so unproductive.' For all his complaints about 'this arid and miserable place', however, Darwin was impressed with the marine environment. 'If we turn from the land to the sea, we shall find the latter as abundantly stocked with living creatures as the former is poorly so.' He was particularly struck by the giant forests of kelp (*Durvillaea antarctica*) that grew on every rock from low-water mark to great depths around the coast and within the channels of the archipelago, and he recalled James Cook's observation that the kelp forests growing off Kerguelen Island spread 'many fathoms on the surface of the sea', their floating leaves acting as breakwaters in exposed harbours. Cook had also noted the kelp's importance as an aid to navigation in the treacherous coastal waters, its presence pinpointing where submerged rocks lay. Darwin wrote, 'I know few things more surprising than to see this plant growing and flourishing amidst those great breakers of the western ocean, which no mass of rock, let it be ever so hard, can long resist'. Even more impressive, he thought, was the abundance of marine life inhabiting the plant. 'I can only compare these great aquatic forests of the Southern Hemisphere with the terrestrial ones in the intertropical regions', he wrote. 'Yet if the latter should be destroyed in any country, I do not believe nearly so many species of animals would perish as, under similar circumstances, would happen with the kelp.'[6]

Kelp is one of the great voyagers of the Southern Ocean. Supported by small chambers of gas, leaves cast themselves adrift from their rock anchors and form floating rafts that cover vast distances, propelled by the complex currents of ocean and air.[7] In 2002 scientists estimated that over 70 million kelp rafts were adrift in the Southern Ocean at any one time.[8] The very large brown species called 'bull kelp' grows in dense forests around many of the subantarctic islands; vast beds of the floating plant end up being jammed into the head of each bay by the unrelenting winds.[9] A study undertaken in 2010 by zoologists from the University of Otago found a myriad of tiny marine invertebrates, including sea spiders, snails and sea stars, sheltering in floating beds of kelp that had washed up on a beach in southern New Zealand. DNA tests revealed that they had migrated from islands hundreds of kilometres away, propelled by the circumpolar winds and currents.[10]

As the *Beagle* sailed through the Le Maire Strait, a group of local Fuegians observing them from a hill began shouting and waving skins, beckoning them to shore. FitzRoy ignored them, so they lit a fire in the pouring rain to signal their welcome.[11] Darwin later described the scene in a letter to his cousin William Darwin Fox, in 1833: 'In Tierra del [*sic*] I first saw bona fide savages; & they are as savage as the most curious person would desire. – A wild man is indeed a miserable animal, but one well worth seeing.' To his mentor Henslow he wrote, 'I shall never forget … They were seated on a rocky point, surrounded by the dark forest of beech; as they drew their arms wildly round their heads & their long hair streaming they seemed the troubled spirits of another world.'[12] While the visitors referred to the indigenous inhabitants generally as Fuegians, there were in fact four different tribal groups, each with its own language and territory. The Selk'nam occupied most of the Isla Grande of Tierra del Fuego, east of the Beagle Channel, while the Haush occupied its southern tip. The Alakaluf inhabited the maze of outer islands along the western side of Tierra del Fuego, using large bark canoes to navigate the waterways in the full force of the Furious Fifties. The Yaghan occupied both shores of Beagle Channel and the cluster of islands bordering the Drake Passage. The early Spanish mariners christened them *indios canoeros* (Indians of the canoes). The German missionary Martin Gusinde called them the 'water nomads of Cape Horn'.[13]

According to oral histories, historical accounts and archaeological evidence, Yaghan people at that time spent much of their lives in canoes, negotiating the intricate channels of the Tierra del Fuego archipelago and relying on its resources. The Argentinian pastor and historian Arnoldo Canclini noted that the Yaghan language included many references to coastal geographical features such as *aia* (bay) and that, unlike the Western notion that land protrudes into the sea, the Yaghan people spoke of the sea entering the land.[14] Dependence on the ocean's resources required excellent navigational skills and mobility, and each family had a canoe with a fire kept lit in the centre. Canoes were made from large pieces of bark stripped from the cherry tree. While the vessels were not suited to the open ocean, archaeological evidence indicates that they did venture across the Le Maire Strait to

Tierra del Fuego showing the location of indigenous peoples in the nineteenth century

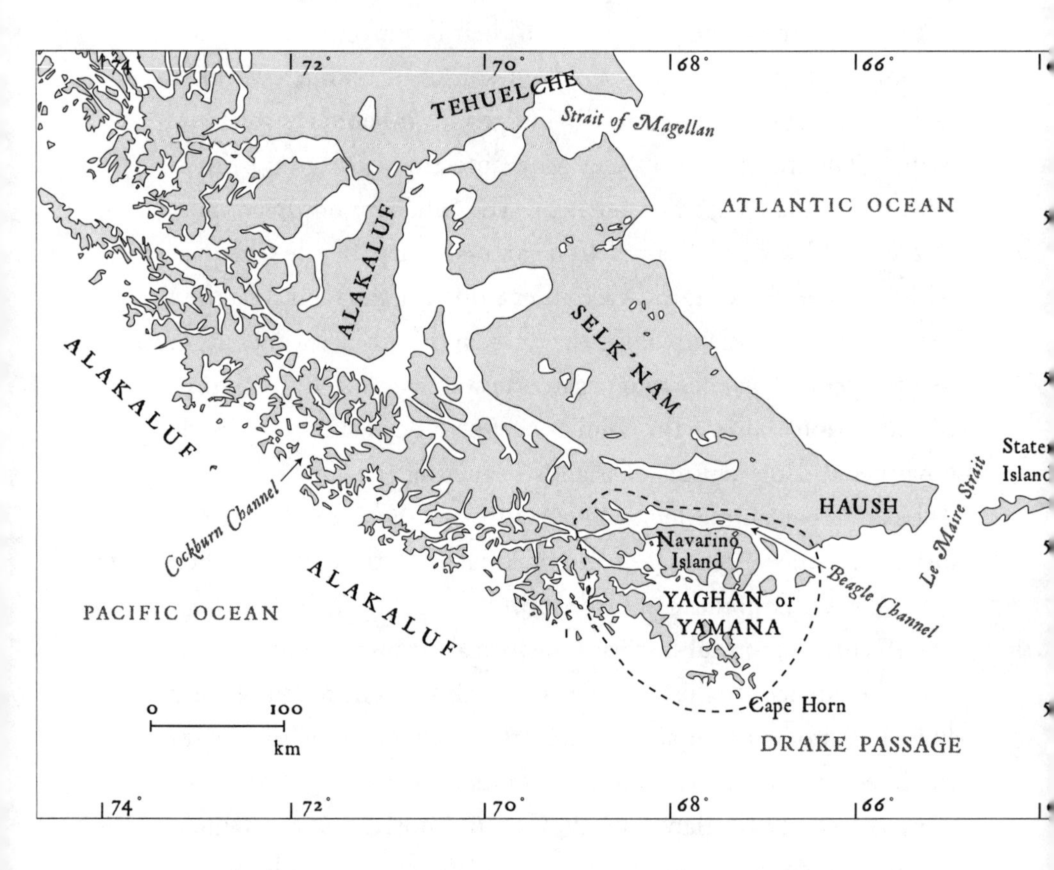

Chuanisin (Land of Plenty), which is now known as Isla de los Estados, or Staten Island. Groups of families shared a particular waterway, with huts providing temporary accommodation along its edges. Large mounds concealing the remains of shell middens associated with these sites can still be seen along the waterways of the Beagle Channel.

The Franco-American ethnographer Anne Chapman spent more than four decades recording the cultural traditions and oral histories of indigenous Fuegians, beginning in 1964 with the last Selk'nam speakers and in 1985 with the Yaghan people.[15] According to Chapman, the Yaghan first encountered Europeans in 1619, when the Spanish García de Nodal expedition entered the Le Maire Strait with the intention of taking possession of the navigable route between the Atlantic and Pacific oceans. Soon after the Nodal visit, they welcomed a group of Dutch mariners who had ventured onto Navarino Island. To the visitors, they were a strong, healthy people distinguished by their long black hair and sharp teeth. The men and women had seal skins draped on their shoulders or around their waists. Some men were painted red and white using ochre mixed with whale or seal oil, which seemed odious to European sensibilities. The Dutch visitors noted with some surprise that they chose to be nearly naked despite the cold climate, but it was their canoes that drew particular comment: skilfully crafted from strips of bark sewn in strips 'giving [them] the shape of a Venetian gondola'.[16]

FitzRoy had explored these waterways during his first voyage on the *Beagle*, in 1829. He accused some of the local people of stealing his whaleboat and attempted to take hostages in retaliation. Most escaped by jumping overboard and swimming for home. He eventually took four captive – two young men, whom he named York Minster and Boat Memory; a boy, Jemmy Button; and a girl, Fuegia Basket – and took them with him when he sailed back to Britain. Charged with missionary zeal, his plan was to 'civilise' them with lessons in English, gardening, husbandry and Christianity before returning them to Tierra del Fuego to become missionaries to their own people. Indeed, FitzRoy's main reason for sailing so far south on his second voyage was to resettle them in their own country.[17] Boat Memory, however, had died of smallpox while in Britain. The other three had attracted considerable interest in FitzRoy's social circles and had been given an audience with the new king, William IV, and Queen Adelaide.

During the second voyage of the *Beagle*, Darwin became acquainted with the remaining three Fuegian captives, observing that they had acute vision and were 'much superior' to any of the sailors on board in being able to distinguish distant objects at sea. He also admired their facility with language.[18] They were skilled navigators, craftsmen and hunters. Nevertheless, in Darwin's view they must have been forced to occupy their inhospitable region by physically stronger and more advanced tribes from the north.[19]

> Whilst beholding these savages, one asks, whence have they come? What could have tempted [them] … to leave the fine regions of the north, to travel down the Cordillera or backbone of America, to invent and build canoes, which are not used by the tribes of Chile, Peru and Brazil, and then to enter on one of the most inhospitable countries within the limits of the globe?[20]

The question of origins and movements of people preoccupied Darwin as much as the origins of plants and animals. Like Johann Forster before him, he believed that human life became increasingly primitive as one travelled southward from the equatorial region.

The idea that the high southern latitudes had a moral dimension gained some adherents during the nineteenth century. William Sollas, a British geologist and anthropologist, outlined his theories in *Ancient Hunters and Their Modern Representatives* published in 1911. He argued that each of the surviving hunter-gather societies, including those encountered by Europeans in Patagonia and Tasmania, represented vanished Palaeolithic, or Stone Age, hunters 'expelled and driven to the uttermost parts of the earth'. 'Justice belongs to the strong', he wrote, 'and has been meted out to each race according to its strengths.'[21]

The American explorer Charles W Furlong spent some time amongst the Fuegian peoples in 1907–8, when *Harper's Magazine* commissioned him to write a story about the Selk'nam people living on the largest island in the archipelago. Furlong became fascinated by the physical and cultural differences between the isolated tribes of the region. By then just 500 Selk'nam people remained, following a campaign of systematic extermination that had begun in the late nineteenth century when settlers claimed their land to graze sheep and cattle. Furlong too favoured the theory that the weaker tribes had been driven southward, forcing the Yaghan to occupy 'that inconceivable labyrinth' at the southernmost point of Tierra del Fuego. But he concluded that the Fuegians were an 'inherently intelligent' and adaptable people, and that there was far less of a gap between 'primitive and civilized man' than others might have argued. In two articles published in *The Geographical Review* in 1917 he explained how the extreme environment had shaped a society in which, apart from occasional gatherings to drive ashore a stranded whale, people generally lived in small, mobile clans or family groups scattered throughout the archipelago. He also observed the complete absence of community leaders. Instead of a formal system of communal decision-making, each canoe represented an independent authority. In this way the clans could not only hold their own amongst competing family groups and neighbouring tribes but 'thrive against those cohorts of nature, storm, cold, dearth of food'. Furlong speculated that the impassable barrier of the Andes mountains had isolated the Fuegians and – paradoxically – protected them from the 'artifices, vices, and diseases of civilization'.[22]

In 1910 Furlong wrote an article for *Harper's Magazine* called 'The vanishing people of the Land of Fire'.[23] By then, colonisation had already sabotaged the Fuegians' fragile existence, but few people were aware of the extent of the destruction of Fuegian society until the 1940s when Francisco Coloane, a local seal hunter, sailor and shepherd, introduced their little-known region to the world through a series of popular novels based on his own experiences.[24] Today the ethnographic collections of missionaries such as Martin Gusinde and Alberto Maria de Agostini, Charles W Furlong's writings and Anne Chapman's exhaustive field research and oral histories document the precious heritage of these southernmost peoples, who have lived in the embrace of the Southern Ocean for over 6000 years.[25]

Even now, a century after Furlong's article, the cultural traditions and environmental knowledge of these remarkable southern peoples survive. In 2014 Will Meadows, a 24-year-old 'young explorer' sponsored by the National Geographic Society, embarked on a journey to Tierra del Fuego as part of a project he called Humanity's Vessel. His aim was to spend time with Yaghan descendants living on Navarino Island and to understand 'what humble sparks still may flicker' from the Land of Fire.[26] He spent a month with Yaghan artist Martin Gonzalez learning the craft of building a traditional canoe from the natural materials found in the surrounding forests and beaches. As he sat on a bench with Gonzalez overlooking Beagle Bay, he contemplated the enduring relationship that the people still have with their islands, and their intimate knowledge of their environment. Gonzalez collected whale bones on certain beaches where whales tended to wash up. He knew how long it would take for the wind and waves to weather the bone until it was hard enough to be used as a harpoon. His ancestors had used such harpoons to spear seals from their bark canoes. In his blog, Meadows reflected on how the Yaghan, far from being extinct, remained in the archipelago 'like the ever persistent Coigue tree in the Patagonia wind'.[27]

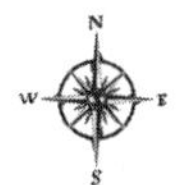

Along with Tasmania, New Zealand and the archipelago that extends from the southern tip of South America, there are around 20 island groups punctuating the vast expanse of the Southern Ocean between latitudes 40° South and 60° South. They are tiny and amongst the most isolated shards of land in the world. The Antarctic lexicographer and historian Bernadette Hince described them as 'the beads of a widely strung necklace around the cold neck of Antarctica, a necklace whose string is water and wind'.[28] But while they share a geographical relationship with the Antarctic continent, the subantarctic islands have their own distinctive physical qualities and human histories. Kerguelen Island, the largest, is part of an archipelago of islets covering 6200 square kilometres roughly halfway between the Cape of Good Hope and Cape Leeuwin, in the path of the Furious Fifties. The McDonald Islands, south of Kerguelen, represent the smallest specks, covering just 2.45 square kilometres. Along with Heard Island, these groups are the highest points of a vast volcanic plateau that lies 4000 metres beneath the surface.[29] To their west, off the coast of South Africa, lie the Crozet Islands, Prince Edward Island and Marion Island. To their north lie Amsterdam Island and Saint Paul Island. Another cluster lies south of Australia and New Zealand: Macquarie Island, The Snares, the Auckland Islands (or Maungahuka) and Campbell Island. Another assemblage forms an arc between South America and the Antarctic Peninsula: South Georgia, and the South Sandwich Islands, South Shetland Islands and South Orkney Islands.

They may be separated by vast expanses of ocean, but they all share the same wild, unforgiving climatic conditions: rain, hail, sleet, snow, unrelenting winds and low temperatures. Most are volcanic in origin and surprisingly young in geological terms.[30] Many are treeless, with tundra vegetation including grasses, sedges and herbs. All host abundant seasonal populations of fur seals, elephant seals, penguins and seabirds, which have adapted to life in or over the Southern Ocean and are directly dependent upon its waters for their survival. Most also depend on finding a spare patch of coast or cliff on which to breed each summer. Fur seals (*Arctocephalus* sp.) feed exclusively at sea, returning to remote beaches each year to give birth, to mate and to moult. Elephant seals (*Mirounga leonina*) were so named because of the large proboscises of the adult males and their habit of roaring during mating season. Other species found in the Southern Ocean are the Weddell seal, which breeds exclusively on the fast ice which floats on the water but is attached to the land, and the crabeater, leopard and Ross seals, which inhabit sea ice (also referred to as 'pack ice').[31] Thanks to human visitors to the Southern Ocean's far-flung islands, a variety of feral animals have also flourished, ranging from pigs and cats to sheep, rabbits and rats – some released intentionally to provide sustenance for stranded humans or to begin farming ventures that were invariably abandoned as a result of the hostile conditions.

The subantarctic islands share a similar pattern of human history, beginning in the early nineteenth century with sealers and whalers and followed by explorers on sporadic visits to learn about their natural history. These were interspersed with occasional shipwrecked survivors seeking refuge. By the twentieth century strategic and economic interests in the region saw the islands divided into territories, as nations from both the Northern Hemisphere (Britain, France and Norway) and the Southern Hemisphere (Australia, New Zealand and South Africa) claimed them as southern outposts to accommodate resupply stations, short-lived commercial enterprises and scientific research; as potential test sites for nuclear weapons; and, more recently, as tourist destinations.[32] As the Tasmanian environmental historian Erica Nathan wrote, 'Attaching nationality to an essentially uninhabitable, geographical outlier entails a dedicated strategic mindset, and perhaps because of this, the polar south is marginal to our identity.'[33]

In 1776 when Cook, en route to the Pacific Ocean, anchored in Bird Bay (which he called Christmas Harbour) in the Kerguelen archipelago, he found plentiful penguins and fur seals that were so tame that his men were able to take as many as they pleased.[34] Animal fur was considered a luxury in the clothing markets of Canton (Guangzhou), London and New York, and fur seals were in demand. The skins were more difficult to prepare than those of otters and beavers, but the Chinese had managed to perfect a technique sometime before 1750, removing the long outer 'guard' hairs to reveal the soft underfur. Groups of sealers began making their way into the subantarctic region in the wake of Cook's reports. Locating just one of these tiny island groups in a smallish vessel in the midst of the Roaring Forties and Furious Fifties – even with reasonably reliable charts and navigational systems – defies the imagination, but sealers were tenacious when it came to finding their prey.

The lucrative fur seal trade transformed the shores of the remote subantarctic islands into theatres of slaughter. Reports of fur seal colonies attracted a flurry of sealing vessels, and a sealing gang would land and remain until they had killed every animal in sight. Sealing captains often dispatched a group of men to spend weeks or months on an island. They would camp in caves or construct crude beach shelters from stone and ships' timbers or ribs from whale skeletons, covering them with tarpaulins or seal skins to keep out the worst of the weather. Researchers have identified the remains of about 50 such shelters in the South Shetlands alone.[35] The British historian of the polar regions Robert Headland described how skinning was done as soon as the animal was dead and before it stiffened or decomposed, which made the process more difficult. The prepared skins were washed in sea water and laid out to dry before being salted to prevent them from turning putrid in the hold of the ship as it passed through the tropics. Headland noted that sealing vessels would land up to 30 men to undertake the work and that a skilled sealer could prepare about 30 skins a day.[36]

Within eight years of the discovery of the South Shetlands by William Smith, the captain of the British vessel *Williams*, an estimated 144 sealing gangs converged on the islands, sailing from New England, Britain and the Australian colonies. Most were there for the first three summer seasons, taking at least 300 000 (and perhaps as many as 900 000) fur seal skins. Some archaeologists have speculated that these sealers were the first humans to over-winter in the Antarctic region.[37]

Fabian Gottlieb Thaddeus von Bellingshausen (whose name in Russian was Faddei Faddeevich, or Faddey Faddeyevich, Bellinsgauzen), was an officer with the Imperial Russian Navy who led the first Russian voyage to the Antarctic region in 1819–21. At the age of 20, he sailed to South Georgia and circumnavigated the South Sandwich Islands before resolving to head for the South Shetland Islands after hearing about sealing activities in the area. There he met William Smith, who boasted to Bellingshausen that he had killed 60 000 seals, while the sealers as a whole had taken 80 000. Bellingshausen predicted that the southern seal colonies would soon be annihilated, observing that there 'could be no doubt' that those around the South Shetlands would be destroyed just as they had been around South Georgia and Macquarie Island. On New Zealand's South Island, 80 000 fur seal pelts were taken in 1824 alone.[38]

The *Chanticleer*, a Royal Navy survey ship, visited South Georgia and the South Shetland Islands during a British scientific expedition in the southern oceans between 1828 and 1831, under the command of Henry Foster.[39] WHB Webster, the ship's surgeon, wrote in his account of the voyage that the islands of South Shetland, including Deception Island, 'formerly abounded with seals'. He continued, 'But such is the havoc made by sealers among them, that they are now scarce and seldom seen. The shores of this basin must have formed a delightful retreat for these persecuted creatures before it was found out by man.'[40]

Sealers operated with impunity amongst the remote seal colonies of the Southern Ocean. Their activities were opportunistic and secretive, usually beyond the scrutiny of metropolitan authorities. Archaeologists and historians are still uncovering the physical remains of sealing activities on the subantarctic islands, reconstructing the lives and activities of sealers and gaining new insights into the nature of their industry.[41] Prior to the arrival of intensive sealing, southern fur seals were estimated to number anywhere between 1 and 2 million. By the time the fur sealing industry collapsed, only a few hundred individuals remained.[42]

Southern elephant seals feed at sea and return to the beaches of the subantarctic islands twice a year – once between September and November to breed, and once between January and April to moult their hair and skin. Their colonies tend to be divided into three distinct groups centred on South Georgia, Kerguelen Island and Macquarie Island respectively. They are extremely capable swimmers, able to cover long distances and dive to depths of more than 1500 metres for up to two hours at a time. Australian scientists have tracked individual seals from the Kerguelen Islands to the Antarctic continent, a distance of several thousand kilometres. It is a vastly different story on land, where the large stores of blubber that sustain the seals during the breeding season render them clumsy and slow, and particularly vulnerable to human hunters.

Sealers turned their attentions to the blubber-rich elephant seals to meet the increasing demand for lighting oil, machine lubrication, paint and soap; the blubber was also used in treating leather and rope and softening wool for clothing. Once extracted, elephant seal oil was odourless and could be stored without going rancid. For all these fine qualities, however, the process of slaughter and oil extraction was a loathsome business. The practice of 'elephanting' generally involved groups of up to 30 men landing on a beach amidst a colony of seals. Armed with clubs, they would first stun the animal then lance it through the heart. The body was stripped of its copious blubber in a process known as 'flensing', and the blubber would then be cut into pieces to be melted in try-pots mounted on a vessel or set up on the beach.[43] The process was similar to that used by whalers. Indeed, the two industries co-mingled in the region, both concerned with the oil-rich blubber that equipped the marine mammals to survive in the high southern latitudes.[44]

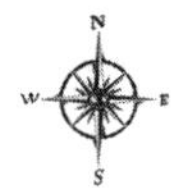

Apart from spasmodic visits from itinerant sealers and occasional shipwrecked sailors, the subantarctic islands also became ports of call for explorers and naturalists, who might spend days or weeks surveying their coastlines and collecting rocks, plants and animals. Johann Forster's observations and specimens of the region's natural history collected during his voyage with Cook, in the 1770s, inspired speculation about the close kinship between some flora and fauna on the Antarctic continent and subantarctic islands despite being geographically separated across vast stretches of ocean. The intriguing questions were how and why certain species migrated between the separate landmasses, and various theories were put forward.

The British botanist Joseph Hooker, who sailed with James Clark Ross on an Antarctic voyage in 1839–43, studied the plants of Kerguelen Island. He noted that the island supported only 18 species of flowering plants, far fewer than the numbers found at similar latitudes in the Northern Hemisphere, but that the vegetation bore a striking similarity to that on the far-distant Falkland Islands. Hooker speculated that the southern biota were so similar that they must have once been part of a great southern continent. Charles Darwin suggested that the seeds of plants found on Kerguelen Island had been carried by icebergs, while the naturalist Henry Moseley proposed that an ancient land bridge had once existed between the subantarctic islands of Kerguelen, Marion and the Crozets, enabling plants to spread across the islands before the land bridge eventually sank beneath the ocean.[45] In 1845, when Hooker presented Darwin with a copy of his *Flora Antarctica*, Darwin acknowledged that the puzzle of how organisms were distributed across the globe would be 'the key which will unlock the mystery of species'.[46] Noting the similarities between the flora of southwestern Australia and South Africa's Cape of Good Hope, Darwin proposed that the remote islands that lay between them could have been parts of a connecting corridor that had enabled plant migration in preglacial times. In a letter to his mentor Charles Lyell in 1857, Darwin emphasised the need for a closer study of the natural history of the isolated southern islands in order to solve the mystery of plant distribution. 'It is my most deliberate conviction that nothing would more aid Natural History, than careful collecting and investigating of ALL THE PRODUCTIONS of the most isolated islands, especially of the Southern Hemisphere … Every sea-shell and insect and plant is of value from such spots.' Saint Paul and Amsterdam islands, he went on, 'would be glorious, botanically, and geologically', while Kerguelen promised to reveal much about past glaciation.[47]

Kerguelen Island is the largest island in the Kerguelen archipelago, which is part of the French Southern and Antarctic Lands. The group lies in the path of the Furious Fifties, at latitude 49° South, and is one of the most isolated places in the world. Along with its nearest neighbours, Heard and McDonald Islands, it forms the exposed parts of a huge plateau that is otherwise underwater and thought to be the largest plateau of its kind in the world.[48]

The islands became the focus of scientific interest in the 1870s when British, French, German and American expeditions landed there, as well as on Campbell and Auckland islands, south of New Zealand, in order to observe the transit of Venus.[49]

In 1874 the British *Challenger* expedition visited the Kerguelen Islands and Heard Island as part of its four-year circumpolar oceanographic voyage of 1872–6, although Joseph Hooker's description of the subantarctic islands in this part of the Southern Ocean could hardly have been an inspiration. Hooker, commenting on the impoverished vegetation of the islands wrote, 'The three small archipelagos of Kerguelen Island (including the Heard Islands), Marion and Prince Edward's Islands, and the Crozets, are individually and collectively the most barren tracts on the globe, whether in their own latitude or in a higher one, except such as lie within the Antarctic Circle itself.'[50] As the *Challenger* tracked far to the south of the Cape of Good Hope, the scientists and crew felt the chill of the westerly winds and the constant pounding of the Southern Ocean. In these conditions, they could not help but wonder at Cook's remarkable voyage to the high southern latitudes a century earlier, in nothing more than 'a little cockshell' as Herbert Swire, a sublieutenant on the *Challenger*, put it.[51] George Strong Nares, the ship's captain, initially tried to make landing on Prince Edward Island but abandoned his plan when a gale blew up and a heavy yellow fog descended. He managed to make landing on Possession Island, one of the Crozets, but found only a lifeless primitive sealers' encampment. Finally he found a more suitable anchorage off the Kerguelen Islands.

When Cook visited Kerguelen in December 1776, during his third voyage to the Southern Hemisphere, he considered that the island deserved the name Desolation given its general lack of vegetation and animals apart from grasses and low-lying plants. 'Perhaps no place, hitherto discovered in either hemisphere, under the same parallel of latitude, affords so scanty a field for the naturalist as this barren spot.' Aboard the *Challenger*, the young crew member Joseph Matkin understood why Cook had been tempted to name the island Desolation: 'Not a tree or shrub was to be seen anywhere, no animals in any sort, neither insects on the earth though we looked carefully, except wild ducks and carrion hawks, we saw no birds, so that we may call it truly a land of desolation.' Lord George Campbell, another sublieutenant on the *Challenger*, was also unimpressed. It was, he wrote, 'a gloomy-looking land, certainly, with its high, black, fringing cliffs, patches of snow on the higher reaches of the dark-coloured mountains, and a grey sea, fretted with white horses, surrounding it'.[52]

Nevertheless, amongst the vast bogs that occupied Kerguelen Island's interior, Cook's surgeon and naturalist William Anderson discovered a cabbage-like plant, now called the Kerguelen cabbage, which was later found to grow on several of the subantarctic islands and proved to be a valuable addition to the diet of sailors on long-distance voyages in the high southern latitudes. Its highly pungent oil is rich in vitamin C, which guards against scurvy. And although there was little sign of life inland, the beaches of Kerguelen Island were teaming with seals and penguins. First American then British sealers had plundered Kerguelen's fur seal colonies since the 1790s. In less than three decades, they had reputedly taken more than a million pelts. Elephant seals and whales also frequented the waters around the islands, their oil fetching good prices in Europe and North America. Fur sealers may not have had the equipment necessary for whaling, but it was not uncommon for whalers to include elephant seals in their catches, and vessels involved in elephant sealing were often counted in surveys of whale fisheries.[53] Cook also remarked on the ease with which the island's penguins could be killed with bare hands; their fate was to be sealed with the same wretched inevitability. The penguins, while far smaller than their mammalian companions, were boiled down for oil that supplied stoves and lamps. According to Moseley, penguin skins were also taken to the Cape of Good Hope, where they were made into rugs and mats.[54]

As the *Challenger* dropped anchor at Kerguelen's main island, Matkin spotted seven or eight gravestones. They marked the solitary remains of whalers who had drowned or died on the island in previous decades. The beaches too were graveyards – littered with massive whale bones, the remains of try-pots and a ship's copper used for boiling down blubber.[55] Over the next three weeks the *Challenger* scientists were kept busy surveying and mapping the island, at least as far inland as the bogs would allow. The Royal Society, recognising the difficulty inherent in undertaking expeditions in the remote subantarctic islands, had instructed them to pay special attention to the botany and zoology of the islands southeast of the Cape of Good Hope, as well as those of Macquarie, Auckland and Campbell islands to the east, and to include the 'nearly unknown' marine fauna in the surrounding seas.[56] The crew made good use of the abundant penguins, seals and ducks to restock their larder, while the expedition's officers located a suitable site for an observatory from which the transit of Venus could be observed later that year by expeditions from Britain, the United States and Germany. In 1893 France officially annexed the Kerguelen Islands, together with the islands of Amsterdam and Saint Paul and the Crozet archipelago.

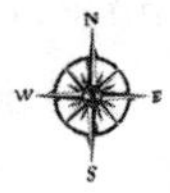

Macquarie Island is a rare place on Earth. It is the only island created entirely from oceanic crust and rock squeezed up from deep within Earth's mantle, and it offers a unique geological window into the planet's deep past.[57] It is also an ecological wonderland where, like its subantarctic island neighbours, huge numbers of Southern Ocean birds and mammals seek refuge, feeding, breeding and nursing their young on its isolated shores.

The island's human history began with the same brutal assault that characterised all of the subantarctic islands. The Sydney sealing brig *Perseverance* is thought to have been the first vessel to report seeing the island, in 1810. Its captain, Frederick Hasselburgh, named the island after Lachlan Macquarie, who was the governor of the colony of New South Wales at the time. Within a decade or so, as reports spread of the island's fur and elephant seal populations, sealing gangs converged and the seal colonies soon succumbed to their frenzied onslaught. The island's environment quickly gained a reputation for being so severe and inhospitable that it was even deemed unworthy by Australia's colonial administrators for consideration as a penal settlement.[58] Captain Douglass, master of the vessel *Mariner*, described his impressions of the 'mountain on the boisterous bosom of the Southern Ocean' when he stopped to load elephant oil for the return voyage to London in 1822:

> As to the Island … it is the most wretched place of voluntary and slavish exilium that can possibly be conceived: nothing could warrant any civilized creature living on such a spot, were it not the certainty of industry being handsomely

> rewarded; thus far, therefore, the poor sealer, who bids farewell, probably for years, to the comforts of civilized live [sic], enjoys the expectation of ensuring an adequate recompence for all his dreary toils. As to the men employed in the gangs … they appear to be the very refuse of the human species, so abandoned and lost to every sense of moral duty.[59]

It was stormy when the British National Antarctic Expedition's vessel, *Discovery*, dropped anchor off Macquarie Island in November 1901. Edward Wilson, the assistant surgeon and naturalist on Robert Falcon Scott's expedition, was keen to visit the island, so he had bribed the *Discovery*'s pilot with a bottle of liqueur to persuade Scott to take the vessel in to land. Scott obliged, and the *Discovery* was soon steaming through a tangle of kelp to anchor off a low-lying shore of the island below two high mountains whose peaks were hidden in cloud. As Wilson stepped out of the ship's boat, he encountered a large brown seal fast asleep on the beach.

> Out came cameras, hammers, guns, rifles, mauser pistols, clubs, and sketch books, till the poor beast woke up and gazed on us with its saucer-like eyes, and then being dissatisfied with the look of some 20 men and cameras, it opened its mouth to its widest, shewing a very old woman's set of teeth, poor thing, and gave a loud inspiratory sort of roar which startled us all.

The creature lunged in an attempt to frighten the intruders away, but it was soon shot and skinned 'in the interests of science', as Wilson put it. The animals lay around the beach and were so fat with blubber that, despite their terrifying roars and habit of frightening unsuspecting walkers in the long hummocks of tussac grass (*Poa flabellata*), they were too heavy and clumsy on land to flee. According to Wilson, they were there for the taking. When he inspected the few huts and sheds on the shore, he discovered that they had been occupied by whalers a year or so earlier. There was a large collection of bird skins in one hut, as well as evidence of a penguin oil industry that had taken advantage of the large colonies around the huts. The whalers had also abandoned vats of the fine oil and what Wilson imagined had once been a putrid pile of rotting penguin carcasses.[60]

Along with its plentiful stock of seals and penguins, Macquarie Island provided a convenient resupply point for Antarctic-bound vessels, thanks to its location at latitude 54° South, nearly 1500 kilometres southeast of Tasmania and about halfway between New Zealand and Antarctica. Its port grew to be a major hub for ships embarking on and returning from hazardous voyages into the Southern Ocean. With the increasing numbers of passenger and trade ships voyaging from Europe to the distant southern colonies of Australia and New Zealand via the great circle route, south of the Cape of Good Hope, Australia's territorial interests in Macquarie Island gained momentum; they were eventually realised when the Australian Antarctic explorer and geologist Douglas Mawson established the first scientific station there during the Australasian Antarctic Expedition in 1911–14. On his arrival at Macquarie Island, Mawson observed what was clearly a thriving penguin oil business. It was being run out of Invercargill on New Zealand's southern coast by Joseph Hatch, a British chemist who had migrated to New Zealand in 1862 and later served as a member of the New Zealand Parliament in the 1880s. He was sending a party to the island each year to kill penguins and elephant seals for their oils.[61]

Hatch had begun his ventures in animal processing with bone milling, rabbit skin exporting and soap and glycerine manufacturing, before turning his hand to fur sealing on New Zealand's sub-antarctic islands. The New Zealand Government had introduced closed periods for sealing in 1873, in an effort to halt the rapid decline in the animals' numbers. After seal skins were discovered on board one of his vessels outside the sealing season, Hatch relocated his operations to Macquarie Island, which was administered by Tasmania and had no such restrictions. There, he developed his business processing elephant seals and experimenting with extracting the oil of king penguins (*Aptenodytes patagonicus*) using a digester works.[62]

By the 1890s, however, newspaper reports that Hatch's sealers were herding live penguins into the digesters had drawn public outcry in Australia and Britain, and Hatch turned to public lectures to promote his venture. During one of these, at the Princess's Theatre in Invercargill, he made the case for his enterprise with a presentation titled 'The Macquaries: beyond the reach of civilisation'. After presenting images displaying the island's beauty, he showed the millions of king penguins that gathered on the island to breed, ridiculing the idea that his operations were causing them to become extinct, since many more chicks fell victim to sea hawks and nellies than to his harvesting activities.[63] Hatch continued to promote the natural beauty and resources of Macquarie Island in public lectures after moving to Hobart in 1912, noting that the people of Tasmania had taken little interest in the island.[64] By then, Hatch was leasing the entire island from the Tasmanian Government, and this was when Mawson's team observed his oil-harvesting operations at first hand. Leslie Blake, an Australian surveyor and geologist with Mawson, observed how about 2000 birds at a time would be driven into a netted enclosure, where a couple of men would pick out the fat one-year-old chicks and knock them on the head before packing them into huge steam boilers to extract their oil.[65] The public perception that penguin oiling was a cruel business never dissipated, and Hatch was still defending himself at public meetings, and campaigning for election to the Tasmanian Parliament, in 1922 at the age of 85.[66] Mawson returned to Macquarie Island with the British, Australian and New Zealand Antarctic Research Expedition (BANZARE) in 1929–30 and was successful in persuading the Tasmanian Government to declare the island a bird and animal sanctuary in 1933, at which point all sealing licences were finally revoked.[67]

Two decades later, in 1959, the Australian marine biologist Isobel Bennett visited Macquarie Island. As mentioned earlier, she was travelling with Mary Gillham, Susan Ingham and Hope Macpherson, on the annual relief and resupply mission for the personnel based there. As they were the first women permitted to travel on an ANARE voyage, their behaviour was closely scrutinised. As Bennett recalled, 'We were invaders in a man's realm and were regarded with some suspicion. We had been warned that on our behaviour rested the future of our sex with regard to ANARE voyages, an attitude which did not particularly amuse us.' As the chartered naval ship approached the island Bennett wrote, 'First impressions are the ones which always remain … high forbidding cliffs looming eerily out of the misty dusk, with an echoing shout of human voices borne on the wind' – the island's station staff welcoming the arrival of the relief vessel – after 'long, dreary months'. On stepping ashore, she could see the signs of the 'wanton and unrestricted destruction of native species' that characterised the human history of the island.[68]

By the time of Bennett's visit there were few fur seals to be seen, and those that did land seemed to remember and showed an innate fear of humans. The elephant seals, on the other hand, dominated the island's beaches in summer months, with up to 50 000 cows coming ashore to give birth. They showed no fear of human visitors and lay about the station between the huts. Even in 1959 scientific understanding of the island's fauna, flora and geological history was fragmentary, but it was clear that its natural and cultural histories were intimately connected with its abundant summer colonies of seals and penguins. Bennett had been sifting through archival records for the history of sealing and whaling in the southern oceans, so she was perhaps more familiar than most with the impact of the industries on the remote subantarctic islands. In 1951 the ANARE scientists had begun a long-term study of the seals' seasonal habits and life histories by branding some of the pups. Branding was, according to Bennett, amongst the most arduous tasks undertaken at the station during the year.[69]

Bennett had been working in Sydney as part of a team led by British-born zoologist William Dakin, examining plants and animals on ocean rock-platforms along the southeastern coastline of Australia. They had noticed that as they went further south, into colder waters, some creatures disappeared and were replaced by others. Like Hooker and Darwin before her, Bennett was fascinated by the geographical distribution of marine life in the Southern Hemisphere and was keen to see whether there were related species inhabiting the colder waters of the Southern Ocean. In her book about her experiences on Macquarie Island, Bennett pondered how Darwin might have regarded the scientific work now being undertaken on the 'desolate, wind-blasted island of mist'. She regarded the subantarctic islands, with their forbidding climate and splendid isolation, as the perfect field laboratories for testing Darwin's evolutionary theories.[70]

Bennett's particular interest on Macquarie Island was in exploring the complex interactions between marine life and the shoreline, and she brought to the subantarctic islands an environmental sensibility first championed by Mawson himself.[71] She ended her book with a warning about the remote island's vulnerability to 'man's ruthless predation'.

> The Island's seashores are far away from the normal haunts of man. It may be that while Macquarie remains a sanctuary, the animals living there will be left to live out their lives as they had done for thousands of years before man's arrival. But with the enormous problems facing an ever-growing world population, with its increasing demands for more and more food, future generations may again take toll from this far-off land.[72]

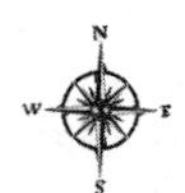

The geographical and scientific work undertaken on the subantarctic islands became an essential part of Australia's claim for their territorial annexation in the years following World War II. Britain had formally claimed possession of Heard Island and the nearby McDonald Islands in 1910, and it was in response to a potential US claim on the islands that the British and Australian governments had hastily convened the ANARE to co-ordinate the establishment of scientific research stations there.[73] The islands lie roughly halfway between Africa and Antarctica. The tiny Heard

Island measures just 40 kilometres from west to east and is dominated by Mawson Peak, an active volcano known as Big Ben. Most of the island is covered by glaciers. McDonald Island, the main island in the McDonalds group, is even smaller. It also has its own active volcano which began erupting in 1992, reshaping the island and doubling its size.[74] The French territory of the Kerguelen Islands, 450 kilometres to the northwest, is their nearest terrestrial neighbour. But while the Kerguelens have a long history of human encounter, Heard and McDonald have had few human visitors: the Australian Antarctic Division estimated in 2005 that Heard Island had received 240 shore-based visits since 1855 and McDonald Island just two.[75] Heard Island was first sighted in 1852 by an American sailor, John Heard, who bestowed his own name upon it. William McDonald did likewise when he discovered the smaller island a year later, and sealers found their way there by the 1880s. The first ANARE headed for Heard Island in December 1947 to make sure that the Australian flag was flying on those tiny, windswept dots in the Southern Ocean, 4000 kilometres from Cape Leeuwin.[76]

Arthur Scholes was a radio operator with that first ANARE to the island, in 1947. In spite of Heard Island's isolation, or perhaps because of it, Scholes recalled being struck by its great beauty. It was, he wrote in his memoirs, 'like a great white iceberg' looming out of a calm blue sea. But first impressions can be deceiving. Within days Scholes considered it 'a depressing place'. He wrote, 'There was little beauty in the gaunt grey rocks, the barren flat and grim precipitous coastline. In the days to come the island's air of sullen harshness was to become all too familiar. But, despite all that, there was something of almost indefinable loveliness about it.'[77]

The ANARE, led by Australian group captain Stuart Campbell, spent a year mapping the island's features, establishing a meteorological station and recording observations of tides, geology and glacial history. A winter in the bleak beauty of Heard Island contained all the right ingredients for Scholes' autobiographical account of the group of 14 men who, with limited resources, succeeded in achieving their scientific mission in one of the most isolated, windy and unforgiving environments on Earth. They arrived with a rough map compiled by sealers in the nineteenth century, and information from four previous expeditions that had briefly visited Atlas Cove on the island's northern coast. A week into their stay, their amphibious aircraft was wrecked by wild winds as they attempted to take aerial photographs of Big Ben.

Between 1947 and 1971 successive ANARE expeditions circumnavigated Heard Island, climbed Big Ben and made the first recorded landing on McDonald Island as well as undertaking seismic, meteorological, glacial and wildlife observations. In many ways, the ANARE was a final salute to the 'heroic age' of Antarctic exploration which had reached its peak in the early years of the twentieth century. After witnessing a flock of skuas attacking and pecking the eyes out of a small seal, Scholes reflected on the island's 'grim prehistoric life in which there was no room for the halt and maimed. Truly, only the fit survive in the Antarctic.'[78]

4

ICE

In all Nature's realm there are few sights more impressive than a vast field of magnificent glittering ice-floes on a beautifully calm morning with the deep blue Antarctic sky overhead. Lonesome, and unspeakably desolate it is, but with a character and a fascination all its own.

Louis Bernacchi, *To the South Polar Regions*, 1901[1]

Latitude 59° 27′ South, Longitude 46° 35′ West: between South Georgia and the South Shetland Islands (6 November 2017)

In the middle of the Southern Ocean, time is measured in latitude and longitude, wave height and wind speed and the proximity of an iceberg. I am feeling disoriented in this ocean, as though time and space have abandoned me. As I look westward, I imagine that I can see all the way around the world. There is no land at this latitude, so I scan the horizon, trying to glimpse the edge of the world. Beyond the limits of my vision are centuries of voyaging and hard labour, letters composed, observations meticulously recorded, measurements made, samples retrieved and minds focused southward, straining to see the first iceberg. Rime ice coats the anchor and deck, and we see our first ice island, with a raft of chinstrap penguins hitching a ride.

In three days' time we will reach Cierva Cove on the Antarctic Peninsula, our farthest south at latitude 64° 08′ South. I will not reach the Frigid Seventies on this voyage. Even so, I am surrounded by ice. 'Ten days after leaving New Zealand', wrote Herbert Ponting, the photographer with Robert Falcon Scott's *Terra Nova* expedition, 'we felt the breath of frozen seas'.[2] Growlers are the modern ship's main adversary these days. They are smaller pieces of ice sometimes called 'bergy bits', and they are masters of disguise, hiding in ocean swells and fog, invisible even to the eye of the radar. Our ship moves slowly. We pass giant tabular icebergs. They have their architectural charms and translucent blue hues, but it is the sea ice that takes my breath away. Sea ice is neither land nor sea but something else entirely. Sea ice is the littoral, the shoreline, of Antarctica. It dictates everything here – movement, temperature, colour, life and death – screeching and grinding and screaming in protest as the ship's strengthened hull forces through it a narrow path.

'To enter Greater Antarctica', the historian Stephen Pyne wrote, 'is to be drawn into a slow maelstrom of ice. Ice is the beginning of Antarctica and ice is its end … Ice creates more ice, and ice defines ice. Everything else is suppressed.'[3] All living things that inhabit the cold polar waters have to survive in ice, and all who voyage to the Antarctic first need to navigate through the sea ice, or pack ice, that surrounds the frozen continent. The sea ice is a vast landscape in its own right, formed from small pieces of ice freezing together to form large masses floating on the surface of the ocean. Antarctic sea ice is highly variable. In winter it is the most distinctive feature of the Southern Ocean, covering 17 to 20 million square kilometres – more than half – of the ocean's surface. In summer it all but melts away to around 3 to 4 million square kilometres trapped within the curving coastline of the Weddell Sea. The former director of the British Antarctic Survey, Richard Laws, called it the 'largest seasonal physical process in the world ocean'.[4] The summer thaw releases flat-topped icefloes and calving icebergs as big as mountains which are destined to drift around on ocean currents before they too melt. Some icebergs travel as far north as latitude 40° South before meeting warmer, equatorial waters. Occasionally, an entire iceberg overturns as it loses its equilibrium, revealing ancient wind-blown, frozen sediments sandwiched in layers of deep blue ice.[5]

According to the people of Rarotonga in the Cook Islands, in about 650 CE the Polynesian explorer Ui-Te-Rangiora navigated *Te-Ivi-a-Atea*, his *waka* (canoe), from Fiji, venturing far enough south to witness the strangeness of the Southern Ocean; he saw bare white rocks rising from the sea to the sky and white powder on the cold waters, and he saw a deceitful animal from the depths of the ocean and the hair of a 'woman of the sea' floating on the waves. Ui-Te-Rangiora is thought to have been describing icebergs and icefloes, elephant seals and floating kelp, and the myth suggests that he may have sailed as far as the subantarctic island of Auckland or the Antipodes Islands.[6] It seems that he was not the only Polynesian to venture into the region before the first

European ships arrived. In 2003 New Zealand archaeologists found the remains of two campsites on an island in the Aucklands, with *hangi* (earth ovens) that date back to between the twelfth and fourteenth centuries containing the bones of sea lions, fur seals and seabirds and the remains of shell fish.[7]

As we have seen, the high southern latitudes were shrouded as well in a Western legend – the Great Southern Land – and it fell to James Cook to prove or disprove its existence when he sailed into the sea ice for a second time, in 1772–5. Cook was not keen on sea ice. He noted in his journal that he would have preferred to sail amongst icebergs in thick fog than to become 'entangled with immense fields of ice'. He did manage to sail to latitude 71° 10′ South without becoming entangled in sea ice, but there was no southern landmass in sight, and he correctly assumed that the enormous 'ice islands' that passed perilously close to his ship had broken away from some icy land further south.[8]

When James Clark Ross entered the sea ice with his fortified sailing vessels *Erebus* and *Terror* during his expedition to the South Pole of 1839–43, he traced along narrow channels, or 'leads', and reached open sea before being confronted, in 1841, by what he called the Great Ice Barrier. He penetrated it, reaching latitude 69° South and the barrier was renamed the Ross Ice Shelf in recognition of his achievement. It is the largest body of floating ice in the world and is located at the southern entrance to the Ross Sea. It extends into the interior of the Antarctic continent, making it an important feature for many of the early expeditions. Ross tried twice more to reach the South Pole, but ice foiled each attempt.[9] Ice barriers, or shelves, form from glacier ice streaming slowly from land into the surrounding ocean and floating there while remaining tethered to the land. The shelves expand each season as more ice flows onto them from the continent, and then shrink as icebergs calve off their edges. This glacial heartbeat of the sea ice is the driver of global weather patterns. Roald Amundsen, the Norwegian polar explorer, who led the first expedition to reach the geographic South Pole, paid tribute to Ross's 'heroic deed'. 'With two ponderous craft – regular "tubs" according to our ideas', he wrote in 1912, 'these men sailed right into the heart of the pack, which all previous explorers had regarded as certain death'.[10]

Australian physicist Louis Bernacchi sailed with the Anglo-Norwegian *Southern Cross* expedition in 1898–1900, but witnessed no icebergs heralding the approach of the sea ice – only a dense fog hovering on the horizon, a slight fall in temperature and a dramatic change in the colour of the sea, to 'dull dirty green'. Bernacchi described how the ice seemed to encircle the continent like a 'mighty spell ... as if to guard the treasures locked up within its bosom'.[11]

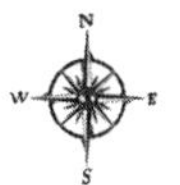

Thirty years after Ross's expedition, the survey vessel *Challenger* sailed into the ice-filled waters of the Southern Ocean on its circumpolar scientific voyage. As the ship approached Heard Island, at latitude 53° South, it met the edge of the Antarctic's summer sea ice. The *Challenger* was ill equipped for ice, but the scientists on board were keen to draw as much information as they could from the ocean at these latitudes. Amongst the *Challenger*'s crew was Joseph Matkin, the young steward's assistant whom we first met in the preceding chapter. He wrote letters home to his cousin John Thomas Swann recounting his experiences of the voyage with the 'scientifics', as the scientists were known to the crew.[12] Like his shipmates, Matkin was in awe when he sighted his first iceberg on 11 February 1874 at 4 am. It measured about 1.5 kilometres in length and 120 metres in height, was square in shape and had snow covering its summit. Over the next few days many more floated near the vessel. On 14 February Matkin counted over 30 large icebergs, each distinctively shaped and coloured: some resembled Gothic cathedrals, with spires and caverns; others haystacks, with snow covering their slopes; and one an iron-clad ship illuminated in beautiful blue tones. There were also masses of ice pieces floating in the wake of the larger icebergs and these, he advised his cousin, would produce ample supplies of fresh water when thawed. At one point chunks of floating ice began to crash and grind against the ship's hull, making a thunderous noise that brought everyone up on deck. Matkin was mesmerised by the sight of the sea ice all around: 'It was a beautiful sight to see the ship ploughing her way thro' it, & the light emitted from the ice made it nearly as light as day. This light is called by polar voyagers, the Ice blink.' The term 'ice blink' describes the effect of light bouncing off a vast field of sea ice in the distance, casting a glare on the underside of low clouds. It creates a celestial map of the ice below, but it also blurs the visible boundary between ocean and atmosphere.[13] A century earlier, ice blink had given Cook his first indication that a field of closely packed ice lay ahead, even as his ship was still sailing in open water. Matkin could just make out a vast barrier of ice suspended in a foggy pale light. At latitude 66° South, within sight of the barrier and surrounded by icebergs, the scientists and crew of the *Challenger* toasted their passage over the Antarctic Circle.

The British Admiralty had forbidden the expedition to venture further than the barrier so George Nares, the captain, turned eastward to search for Termination Land, a feature first recorded in 1840 by Charles Wilkes, leader of the 1838–42 United States Exploring Expedition to the Antarctic.[14] By 20 February the *Challenger* was becalmed. But for the freezing temperatures and 50 or so icebergs, Matkin could almost imagine himself in the quiet waters of the tropics. After taking sounds and sailing under steam for two days, Nares determined that Termination Land must have been a mirage. Matkin reflected on how the *Challenger* crew had also spied what appeared to be land; but, on closer inspection and with the benefit of steampower, it had turned out to be mere clouds of vapour. Ice has a habit of playing tricks with the senses. The next day the *Challenger* clipped an iceberg and sustained minor damage, just as the thermometer dropped and a gale blew up. Matkin recorded the panic on board when another large iceberg suddenly loomed out of the fog beside the ship:

> The confusion was something fearful; nearly everyone was on deck, it was snowing & blowing hard all the time; one officer was yelling out one order, & another something else. The engines were steaming full speed astern, & by hoisting the topsail the ship shot past it in safety. A seaman fell from the trysail while they were hoisting it, & was much hurt.

Icebergs eventually surrounded the ship, and layers of ice blanketed its sides and rigging. At one point a boy fell from the rigging into the ocean and was quickly retrieved, half frozen. The captain decided to take shelter from the wind under the lee of a large iceberg, a manoeuvre that a wind-powered vessel would never have managed. Matkin thought it was the 'worst and most dangerous night' they had experienced. Even as they headed northward, into warmer latitudes, the *Challenger* narrowly avoided hitting another iceberg; they 'shot past it close enough to allow a biscuit to be thrown on to it'. By this point, Matkin confided to his cousin, everyone who had been eager to see icebergs was heartily sick of them.

Matkin may not have cared whether or not he ever saw another iceberg, but by the time of the Sixth International Geographical Congress in London in 1895, the ice of the high southern latitudes was a cause célèbre in scientific circles. Delegates discussed the current state of geographical and scientific knowledge, and they were unanimous in declaring that 'the greatest piece of geographical exploration still to be undertaken' was in the Antarctic, expeditions which would result in 'additions to knowledge in almost every branch of science'.[15] In 1900 the president of the Royal Geographical Society, Sir Clements Markham, addressed a meeting in London urging members to support a British expedition. Polar voyages, he noted, were the best training grounds for teaching members of the British Navy 'self-reliance, quickness of eye, steadiness of nerve, and the necessity of comradeship'.

> Should we not rejoice at giving them an opportunity to do battle with and to conquer the antarctic ice, as our navy always conquers ... It is from the furious gale, off the

> frozen lee shore, among the hardships and perils of polar navigation, that Britons learn those qualities which have made so many enemies quail before our unconquered fleets. Even if there is no gain to science, still it is well that our seamen should defy the obstacles of the frozen sea.[16]

The National Library of Australia holds a rare and special copy of *The Antarctic Manual for the Use of the Expedition of 1901*. The blue-bound book, published by the Royal Geographical Society, is the copy that sailed with Scott's British National Antarctic Expedition in 1901–4, one of the 1200 books that were included in the *Discovery*'s library. The preface, written by Markham, reflects his preoccupation with the naval tradition of polar exploration that would be brought to bear in the Antarctic region. But apart from its value as a resource for the *Discovery* expedition, the book also illustrates just how little was known about the Southern Ocean environment 25 years after the *Challenger* expedition; this slim volume contains the sum of scientific knowledge about the Southern Ocean and the Antarctic continent at that time, covering everything from observations of climate to terrestrial magnetism. The book begins with a glossary of 'ice nomenclature', describing more than 70 different types of ice – many originating from the Arctic region – that the expedition might expect to encounter. This is followed by a set of instructions prepared by 'leading men of science' about the various types of investigations to be pursued; a selection of travel narratives written by earlier explorers in the high southern latitudes; an extensive bibliography compiled by the society's librarian, Hugh Robert Mill; and a set of neatly folded maps, of different scales, showing the tracks of earlier voyages to Antarctica.[17]

The book is a relic of the 'heroic age' of Antarctic exploration, written just as a series of polar explorers were setting forth to transform the remote continent into a stage for scientific discovery and national glory. During this period countries including Britain, Belgium, France, Germany, Norway, Sweden and the newly federated Australia mounted government-sponsored exploratory expeditions to the Antarctic region. Britain's first venture was the *Southern Cross* expedition, of 1898–1900, led by Carsten Borchgrevink; it was followed by the higher profile expeditions led by Robert Falcon Scott and Ernest Shackleton. The Belgian Antarctic Expedition, of 1897–9, was the first to winter in the Antarctic region. The French Antarctic Expedition included two voyages led by Jean-Baptiste Charcot, in 1903–5 and 1908–10. The first German South Polar Expedition was led by Erich von Drygalski, in 1901–3, and the second by Wilhelm Filchner, in 1911–12. The Norwegian Antarctic Expedition led by Roald Amundsen, in 1910–12, was the first to reach the geographic South Pole, while the Swedish Antarctic Expedition, led by Nils Otto Gustaf Nordenskjöld, in 1901–4, spent two seasons in the region. The Australasian Antarctic Expedition, led by Douglas Mawson between 1911 and 1914, included men from Australia, New Zealand, Britain and Switzerland.

The *Discovery* expedition was intended to be scientific in nature, although Markham regarded its primary objective as geographical discovery.[18] Markham appointed John Walter Gregory, professor of geology and mineralogy at the University of Melbourne, to lead the civilian scientists. Gregory was keen to find evidence in the Antarctic that might provide clues as to the nature of ice in the 'Great Ice Age'. A similar sheet of ice in its glacial movement, he speculated, might have shaped Europe's own climate. But Gregory withdrew over Markham's insistence that Scott would have complete command over the scientists on board.[19] He was replaced by George Murray, a British publisher and editor of *The Antarctic Manual*, who had prepared a detailed program of research for the expedition emphasising the close study of icebergs and sea ice during the voyage south. As part of this, the scientists were to inspect the Great Ice Barrier in order to determine its origin, structure and movement.

On 16 November 1901, the expeditioners were on deck scanning the horizon when officer Michael Barne yelled 'Ice!' before slipping over in his excitement. Edward Wilson, the expedition's assistant surgeon and naturalist, recorded in his diary, 'All he had seen was a piece of white ice the size of a soup plate, and yet in ten minutes there were bigger pieces all round us and everyone on the bridge to see it … champagne at once ordered for dinner this evening.' Wilson had also caught a glimpse of a long white streak of white on the horizon. The *Discovery* was soon in sea ice. Everyone on board wanted to taste a piece, finding it quite palatable and much less salty than the surrounding sea water.[20]

The expedition had reached the place where sea and sky merged, and where land ice transitioned to sea ice.[21] As the *Discovery* crunched and grated its passage through the dense pack, Wilson was simply spellbound. Entering the sea ice was a sensory experience as much as a physical one. The ice was 'a perfect miracle of blue and green light', he wrote, 'and then came the ice birds'. In a surge of colour, great flocks of birds escorted the *Discovery*: bluish-grey southern fulmars; Antarctic petrels; pure white, graceful snow petrels; black-and-white pintado petrels; and sooty albatrosses. The expeditioners' first 'real' iceberg was old and worn but impressively proportioned. Wilson noted that it had toppled over to expose stratified layers of blue-and-white and green ice. The icebergs' nature and origin may have been mysteries, but their fantastic shapes and varying hues captured Wilson's artistic imagination: 'The ice only had its whiteness broken with the most exquisitely shaded blues and greens – pure blue, cobalt and pale emerald green and every mixture in between them. I never saw a more perfect colour or toning in all nature.'[22]

While Wilson sketched the icebergs, others measured them and recorded their bulk and appearance. Scott estimated one particularly large example, off King Edward VII Land, to be about 11 kilometres long and 70 metres high. Unlike those in the northern polar region, these southern tabular icebergs appeared to have 'broken quietly away from some huge sheet of ice'.[23] The iceberg, Stephen Pyne wrote, is 'the most complex ice mass', a 'frozen record' of time and movement. 'Its dazzling whiteness masks a dense fabric of acquired ices and shapes.'[24] Far from being a monotonous white, the sea ice encircling the Antarctic continent was suffused with colour. Later in the day, the sun broke through grey clouds 'and burst with astonishing brilliancy and beauty on all those fields of ice'. Wilson was particularly struck by the way the sea ice almost seemed to be a living thing.

> The strangest thing perhaps in the pack is the constant motion, and the gentleness of it, and yet the irresistible force of big masses of ice. The swell was very big, but the surface of the water was merely rippled like a lake, and in places not even rippled, so that the ice was beautifully reflected in the smooth water and yet this incessant immense swell rising and falling, like a breathing in sleep.[25]

As the *Discovery* approached the Great Ice Barrier, Scott sailed back and forth along its face, comparing his route with that plotted by Ross in 1841. He was surprised to find that the ocean edge of the barrier had retreated up to 48 kilometres in the time between the two expeditions. He also noted that, when the ship was tethered to the barrier, the ice rose and fell with the ship.[26]

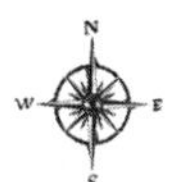

Australia's oldest science and natural history collection lies deep in the chilly sandstone catacombs of the Australian Museum, in Sydney. The collection was established in 1837 with the aim of procuring 'many rare and curious specimens of Natural History', the fruits of colonial occupation and exploration in this rare and curious outpost of empire.[27] It includes creatures from the Antarctic and the Southern Ocean, now residing in the Spirit House with its row upon row of jars containing specimens dredged from the depths, floating eerily in preserving fluid. The house is a labyrinth of rooms, filled with more than 18 million specimens. In the corner of one room 'Shackleton's emperor penguine' stands stiffly within its glass enclosure, peering intently towards an unseen horizon. More than a century after its death in the name of science, it still seems to be looking for the ice of its homeland.

Ice is the language of ocean and land in the high southern latitudes, and it is on the floating platforms of ice, those liminal places between land and ocean, that emperor penguins gather every year to mate and, if conditions are right, to lovingly raise their chick. The largest and heaviest species in the penguin family, the emperor was given its scientific name, *Aptenodytes forsteri*, in honour of Johann Reinhold Forster, the naturalist on board Cook's second voyage to the Southern Ocean. (*Aptenodytes* means 'featherless diver'.) It is likely that Forster was the first person to see the bird, although he mistakenly identified it as a king penguin, the emperor's closest relative.

Like the ice shelves that support their breeding colonies, emperor penguins are creatures of both land and sea. They are well adapted to the extreme cold of an Antarctic winter, with four layers of scale-like feathers. Mature birds, both male and female, stand about 70 centimetres tall, and both are remarkable divers. They gorge on an assortment of fish, krill and squid.[28] By the time they join their colonies to breed, in late autumn, they can weigh up to 40 kilograms. Pairs mate and lay a single egg. As winter descends, the male draws on his body reserves to incubate the egg and raise the chick while the female leaves the colony to feed.

The emperor penguin is the only animal to breed in the extreme cold of the Antarctic winter, and the males must rely on each other to survive. They do so, in a rare and remarkable demonstration, by huddling together in tight groups to share warmth and resist the freezing katabatic winds that blow down from the polar plateau.[29] The birds take it in turns to stand on the outer edge of the group. One by one they peel off the windward side and shuffle along to the leeward side before rejoining the huddle, all the while balancing their precious single eggs on top of their feet. By the time the females return, in July, the males have lost almost half of their total body mass. The chick-rearing season then runs from August to December.[30]

Apart from his duties as assistant surgeon on the *Discovery* expedition, Edward Wilson documented his observations of the natural world in his diary as well as in numerous sketches and paintings. He had recently completed the first comprehensive account of Antarctic and subantarctic penguins, based on brief visits to New Zealand, Macquarie Island and McMurdo Sound, and he developed a particular interest in emperor penguins when he discovered the first-known breeding colony, on sea ice under the cliffs of Cape Crozier, on Ross Island.[31] Finding that the chicks were fledging in December and January, he speculated that they must have hatched from eggs laid during the coldest months of the year, a breeding cycle that was 'eccentric to a degree rarely met with, even in Ornithology'.[32] He made several visits during the expedition, observing that the chicks seemed to have an extraordinarily high mortality rate of about 70 per cent.

Few Antarctic explorers had even seen emperor penguins, much less witnessed their breeding practices, but to naturalists of the day they represented an intriguing evolutionary mystery, and the scientific consensus was that they were a primitive form of bird. The icy, eternally cold Southern Ocean region resonated powerfully in the Victorian imagination, and scientists believed its creatures, like its indigenous peoples, were the living embodiments of 'primitive' societies doomed to extinction.[33] Wilson was aware of an archaeological discovery made in 1861 of the fossilised remains of a small dinosaur, *Archaeopteryx*, in Jurassic-era limestone in Bavaria, Germany.[34] Its features placed it somewhere between a reptile and a bird, and Wilson was keen to examine emperor penguins at different stages of development in the hope that they might reveal their evolutionary secrets. 'The possibility', he wrote in 1907, 'that we have in the emperor penguin the nearest approach to a primitive form not only of a penguin but of a bird, makes the future working out of its embryology a matter of the greatest possible importance.'[35] When Wilson returned to

Antarctica with Scott's *Terra Nova* expedition in 1910, he was keen to test a theory proposed by the German zoologist Ernst Haeckel that the evolution of species was mirrored in their embryonic development. According to the theory embryos of more advanced species, such as humans, passed through stages in which they displayed the adult characteristics of their more primitive ancestors.[36] Wilson decided that studying emperor penguin embryos might reveal an evolutionary link between birds and reptiles.

In the depths of the 1911 winter, Wilson left Scott's base camp with the expedition's assistant zoologist, Apsley Cherry-Garrard, and Henry 'Birdie' Bowers. They trekked 100 kilometres in complete darkness with temperatures reaching as low as –60 degrees Celsius in order to reach the emperor penguin breeding colony at Cape Crozier that Wilson had sighted ten years earlier. There were fewer than 100 penguins – fewer than Wilson had observed on his earlier visit – all huddled together under the barrier cliff and trumpeting loudly as the men approached. As Cherry-Garrard related in his memoir,

> After indescribable effort and hardship we were witnessing a marvel of the natural world, and we were the first and only men who had ever done so; we had within our grasp material which might prove of the utmost importance to science; we were turning theories into facts with every observation we made – and we had but a moment to give.

The men managed to retrieve five live eggs, keeping them inside their mittens for warmth. Two broke on the harrowing journey back to the base camp, but Wilson managed to cut open and remove the embryos from two of the remaining three eggs which he pickled and eventually delivered to the Natural History Museum in London where they have remained. Cherry-Garrard later recounted the six-week trek in his memoir, *The Worst Journey in the World*, describing it as 'the weirdest bird's-nesting expedition that has ever been or ever will be', while Scott assured the disappointed Wilson that their journey across the ice had been 'one of the most gallant stories in Polar History'.[37] Sliced and mounted onto slides, the embryos remained unexamined until 1934. By then, however, the evolutionary theory that inspired the expedition had been rejected. The eggs may have had little scientific importance, but Cherry-Garrard's so-called winter journey continued to feed the British public's appetite for stories of the 'heroic age' of Antarctic exploration.[38]

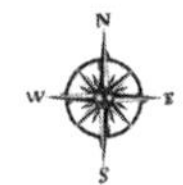

Ill health forced Ernest Shackleton to leave Scott's *Discovery* expedition early, but his ambition to reach the South Pole never subsided and, after gaining the necessary support to lead his own expedition, he embarked upon it on New Year's Day 1907, departing from Lyttelton, New Zealand aboard the *Nimrod*.[39] Fifteen days later, after a turbulent ocean crossing in gale-force winds, the small whaling vessel entered a highway of floating ice. Shackleton wrote in his expedition report,

> We entered the city of the white marvellous Venice of the South – mile upon mile of great icebergs; never a sign of pack ice, but from the crow's-nest on the main mast

> stretched out east and west as far as we could see were these wonderful heralds of the frozen south – great giants weaned from the bosom of Antarctica.[40]

Aboard the *Nimrod* was TW Edgeworth David, otherwise known as 'the Professor', a Welsh-Australian geologist whom Shackleton had invited to join the expedition as chief scientific officer. Like Wilson, he was intrigued by the idea that emperor penguins were 'living fossils'. In his notes taken during the expedition, later serialised in the *Sydney Morning Herald*, David wrote that recent discoveries of fossilised penguin bones in New Zealand and West Antarctica suggested that the emperor penguin was descended from an ancestor living when all the great southern continents were one. As the *Nimrod* anchored on the edge of a mass of thick bay ice, David noticed a group of emperor penguins on it, 'standing upright, about 4ft [1.2 metres] high, on their broad, squat, webbed feet, and apparently semaphoring to one another by alternatively raising and dropping the flippers which constitute their apologies for wings'. As the men watched the penguins, the penguins stood in pairs, like sentries, watching the men. With the bird's 'gentleman's swallow-tail coat' trailing in the snow, David wrote, 'he walks with the certain knowledge that his braces are broken, and that the eyes of the world are upon him'.[41]

In August 1914, as the hostilities of World War I engulfed Europe, Shackleton sailed for the Antarctic once again, aboard *Endurance*, leading an expedition to the ice for the second time. Roald Amundsen and Robert Falcon Scott had already reached the South Pole on separate treks in 1911, with Amundsen reaching the pole first, followed by Scott 34 days later. Scott and his four companions had perished as they attempted to return to their vessel. Shackleton was determined to be the first to trek across the entire continent, although those who signed up for this Imperial Trans-Antarctic Expedition never actually set foot on it. A most unusual Antarctic expedition, it soon became a struggle for existence on the surface of floating sea ice.

The *Endurance* left Grytviken whaling station on South Georgia on 5 December 1914 and, within two days Frank Worsley, the ship's master from New Zealand, was steering through sea ice. Worsley had enthusiastically signed up for his first voyage into the Southern Ocean. He was an experienced navigator and a meticulous man by nature, and his charts, logs and diaries ultimately survived as the only written records of the expedition, providing the raw material for both his and Shackleton's later accounts.[42] Worsley thought the passage through the sea ice, ramming the vessel into icefloes in order to open up cracks to pass through, was 'exciting work'. At the edge of one small floe, he noted three emperor penguins apparently deep in conversation and admiring their own reflections in the ice. Neither the penguins nor the two blue whales that surfaced nearby took the slightest notice of the strange vessel grinding its way along nearby.

On 10 January 1915, Worsley sighted part of the Antarctic continent called Coats Land. The water was 'turgid with diatoms' and the nearby sea ice was crowded with an assortment of crabeater seals, sea birds and Adélie and emperor penguins. This was a polynya, an area of open water, that formed and re-formed annually along the Weddell Sea coast as a result of strong katabatic winds. As the ship steamed along the ice cliffs of the sea, the men on board spotted a group of 40 emperor penguin fledglings and captured 11 as food or biological specimens. When they set three or four free, 'the departing birds turned round, gave us a little bow and then hopped over the rail onto the ice, where they again bowed and walked off. It was so extraordinarily human as to be almost uncanny'.

By the week's end, the vessel was caught fast in sea ice and drifting away from the coast. At times, the ice seemed to be alive and waging its own battles, as floes crashed into each other before launching combined attacks on the ship. These were midsummer months, but the temperature fell to –50 degrees Celsius and the sea ice eventually froze into a solid mass around the ship. The *Endurance* drifted in the pack for more than nine months but, on 21 November 1915, it was finally crushed by the slow but irresistible forces of the moving sea ice. An attempt to sledge to land failed, and Shackleton decided that the drifting icefloe christened Patience Camp would be their best chance of reaching safety.

For the next five months, the men experienced all that a drifting icefloe can offer: cracks suddenly opening up beneath their tents, leopard seals lying in wait for a meal of penguin at the edge of the floe, a shortage of fresh water while all around lay a vast desert of sea ice too salty to drink. But in spite of all the dangers in those liminal spaces between ice and water, penguins proved to be their most cherished companions. Then, as winter closed in and the last remaining penguins and seals disappeared, the men were left with the loneliness of the sea ice. 'Our craving', wrote Worsley, 'to see some living, breathing creature, any creature at all, may be imagined when I say that we missed them as though they had been our personal friends.'

The importance of these early encounters with emperor penguins was largely overshadowed by the human dramas of the expeditions. But after making his first voyage to the region in 1946, at the age of 20, British penguin biologist Bernard Stonehouse drew attention to the birds' scientific importance. Even then biologists knew very little about the penguins' range and breeding habits so, when Stonehouse discovered a rookery on the Dion Islands off West Antarctica in 1949, he and two companions decided to spend three months living nearby in tents in order to study them during the winter breeding season.[43] Since then, penguin biologists have identified at least 40 colonies in which emperor penguins breed each year, mostly on the more stable shelves around the Antarctic coastline.[44] Nevertheless, they are considered to be vulnerable to the changing sea ice conditions, and their future is uncertain.[45] A precarious existence indeed.

When early polar explorers began to traverse the icy expanses of the Antarctic continent, they were surprised to see the abundance of life around its edges. Aboard the *Terra Nova* in 1910, Scott observed the myriad creatures that could be found living amidst the sea ice. At the edge of every icefloe there was an assortment of crabeater seals, snow petrels and Adélie penguins feasting on a banquet of krill. Nearer the coastline, the occasional larger mammalian predators patrolled for an easy meal on ice. These included leopard seals, pelagic whales and the 'unappeasably voracious' orcas. In contrast to the barren conditions on the polar ice sheet, the waters around the ice were brimming with life. The sea ice was unusually thick that year, and Scott's Western Party (one of three scientific exploring parties formed for the expedition) spent 20 days pushing through it.[46]

Sea ice is notoriously fickle. The ship might enter an area of thin ice that gave way easily, only to be brought to a halt by a small floe that seemed to Scott to be 'possessed of an evil spirit'. On occasions, caught between two floes, the *Terra Nova* would swing around and fall away, then drift to leeward before the next assault on the ice. Each manoeuvre could take up to half an hour, and Scott found himself being entertained by groups of Adélie penguins diving for food under the ship. He thought them 'wholly ludicrous' on the ice, but underwater their agility and speed were astonishing.[47] Discovered in 1840 by the French Antarctic Expedition led by Jules Dumont d'Urville, Adélie penguins are excellent swimmers as well as being adept on land. They breed on land in the warmer months and spend winter in the sea ice. They are also capable of walking long distances across ice to reach their colonies, occasionally sliding on their bellies where there is sufficient snow covering.[48] During the *Nimrod* expedition a year or so earlier, the Adélies had won over everyone with their human-like antics. David wrote, 'They are the dearest, quaintest, and most winsome birds imaginable.'[49] They would come running up to the men, waving their flippers as if to signal for them to wait, and occasionally tripping over in their haste. Shackleton's expedition artist, George Marston, would mimic their movements then lead them in a procession over the ice, issuing orders as they copied him with a series of military-like manoeuvres that brought roars of laughter and applause from the men and, no doubt, confusion to the penguins. Such interactions between humans and penguins may have been heart-warming in any other context, but in reality most ended with the animals being slaughtered either to supplement the men's diet with fresh meat or to be preserved as specimens. Some of the expeditioners felt the incongruity of the situation. Scott reflected, 'It seems a terrible desecration to come to this quiet spot only to murder its innocent inhabitants and stain the white snow with blood; but necessities are often hideous'.[50] Large numbers of Southern Ocean birds and animals were also routinely collected on such expeditions for dispatch to various museums for further examination and display. Shackleton's *Nimrod* expedition yielded a particularly large collection of natural history specimens, including the skins of emperor penguins which were donated to various museums around the world.[51]

The floating icebergs and icefloes dominated the ocean surface, but beneath them were whole communities of marine life. While the *Endurance* was imprisoned, one of Frank Worsley's daily tasks was to cut holes in the thick icefloes surrounding the vessel to take soundings of the depth of the sea beneath the hull. He used the opportunity to dredge up biological specimens from the seabed for the expedition biologist, Robert Selbie Clark, and their hauls invariably yielded an array of colourful species, from bright red prawns to blue glacial mud dropped from the bottom of icebergs. One day the dredge came up with sponges bearing long, glass-like spicules up to 30 centimetres in length. When the *Endurance* sank, the whole collection was returned to the seabed.[52]

Iron and other nutrients leach from icebergs into the surrounding water, fostering the growth of microscopic plants and algae at the bottom of the ocean food chain. These phytoplankton play a crucial role in removing carbon dioxide from sea water and releasing oxygen back into the atmosphere. Almost a century after Shackleton's *Endurance* expedition Norbert Wu, a renowned underwater photographer, ventured below the Antarctic ice sheet near McMurdo Station to film life in the depths of the Ross Sea. His images revealed a colourful and dynamic world in stark contrast to the ice desert above.[53] As Scott put it in his *Terra Nova* journal, 'As one looks across the barren stretches of the pack, it is sometimes difficult to realise what teeming life exists immediately beneath its surface ... Beneath the placid ice floes and under the calm water pools the old universal warfare is raging incessantly in the struggle for existence.'[54]

5

DEEP

I see always the steady, unremitting, downward drift of materials from above, flake upon flake, layer upon layer – a drift that has continued for hundreds of millions of years, that will go on as long as there are seas and continents. For the sediments are the materials of the most stupendous 'snowfall' the earth has ever seen.

Rachel Carson, *The Sea Around Us*, 1951[1]

Latitude 64° 08′ South, Longitude 51° 00′ West: between Antarctic Sound and Cierva Cove (9 November 2017)

I am mesmerised by the incessant movements of the sea ice as it is stirred by invisible surface currents. In places, clumps of ice bump and crush against each other, forming circles with rolled-up edges like ice flowers. We are sailing through a garden of white lotus blossoms floating on cobalt blue water, and the effect is magical. When we land at Brown Bluff, on the Antarctic Peninsula, I am dazzled by a glorious blue sky and powder snow. The rocky promontory casts a copper glow over a colony of Adélie and gentoo penguins. The ocean here is a deep, soft aquamarine tone. The clarity of the water is startling. Beyond my field of vision, beneath the deep scouring of moving ice, I imagine starfish and anemones, corals and sea butterflies, and all manner of undersea life. This is where the penguins finally reveal themselves to be true birds. They may waddle awkwardly on ice or land, but under the surface they soar with effortless grace and speed and agility. They are the albatrosses of the deep.

Scientific curiosity about the nature of the ocean bed was stimulated during the mid-nineteenth century by plans to lay cables on the floor of the Atlantic Ocean between Europe and North America, and such a feat of engineering required a detailed knowledge of that floor.[2] The maritime nations of Europe and the United States also depended upon reliable maritime charts that plotted safe anchorages and hazards to be avoided along trade and passenger shipping routes crisscrossing the major ocean basins. As the American ocean historian Helen Rozwadowski wrote, sampling and charting the deepest parts of the ocean was becoming an act of national importance and bestowed a certain status upon those involved, in the same way that the discovery of new lands had brought national acclaim to earlier explorers of the new terrestrial world of the Southern Hemisphere.[3]

As for the Southern Ocean, the main incentive to investigate its waters had been to exploit the region's fur seal populations. By the mid-nineteenth century, government sponsorship of three major voyages into the Southern Ocean signalled an emerging interest in exploring the high southern latitudes, although this was driven as much by national rivalry as by scientific curiosity. Each of the expeditions – the 1828–31 British *Chanticleer* expedition to Deception Island in the South Shetland archipelago; the 1838–42 United States Exploring Expedition led by Charles Wilkes; and the 1839–43 British-sponsored expedition led by James Clark Ross – had similar aims. All were primarily interested in studying terrestrial magnetism and its impact on ocean navigation. This was a subject of particular scientific interest during the nineteenth century, culminating in concerted efforts to locate the South Magnetic Pole in the early twentieth century. The campaign to establish the laws of magnetic variation, intensity and dip became known as the Magnetic Crusade.[4] Each expedition was also required to collect specimens of flora and fauna as well as information about the nature of the currents and meteorology of the region. Wilkes's narrative, for example, included a vivid description of the rich marine life in the waters of the high southern latitudes and one of his scientists, James Dana, published the first detailed description of Antarctic krill, drawing on a specimen and notes from the expedition.[5]

However, even towards the end of the nineteenth century, although some scanty information had emerged from earlier surveys of surface and shallower waters associated with coasts and shipping routes to determine opportunities for commercial benefit – particularly new whaling grounds – precious little was known about the Southern Ocean's depths. James Clark Ross had made the first truly deep sea soundings in 1840, with his weighted hemp line reaching an estimated depth of 4427 metres.[6] But the idea prevailed that little or no life could exist in the perpetual darkness and crushing pressure below around 550 metres.[7] Even those with scientific training dismissed the idea that relatively advanced life forms could exist at great depths, arguing that any creature caught by sounding lines was likely to have become entangled as it swam near the surface. In the 1830s the British biologist Edward Forbes had developed the azoic theory, basing it on his experiences of dredging in the depths of the Aegean Sea. He wrote, 'As we descend deeper and deeper in this region, the inhabitants become more and more modified, and fewer and fewer, indicating our approach towards an abyss where life is either extinguished, or exhibits but a few sparks to mark its lingering presence.'[8]

In 1872 the *Challenger*, a three-masted, square-rigged warship, set sail from the English town of Sheerness with the primary aim of investigating the nature of the deep sea. It was the first voyage of its kind, a truly scientific voyage, dedicated to acquiring information and samples from the depths of the world's major oceans. Sailing under the command of George Strong Nares, the vessel carried 20 officers, 200 crew members and a team of 6 civilian scientists led by the professor of natural history at the University of Edinburgh, Charles Wyville Thomson. The expedition was required

> to investigate the physical conditions of the deep sea throughout the three great ocean basins, that is, to ascertain their depth, temperature, circulation, &c., to examine the physical and chemical characters of their deposits, and to determine the distribution of organic life throughout the areas traversed, at the surface, at intermediate depths, and especially at the deep ocean bottoms.[9]

The London Society had commissioned the vessel for the voyage and replaced its 17 guns with natural history and chemistry laboratories, as well as dredging equipment and specialised instruments for measuring current flow and air and water temperature. While the ship relied on sails for most of its time at sea, it also had a small steam-powered engine capable of producing over 1200 horsepower, which proved invaluable for the dredging operations and negotiating the treacherous sea ice of the Southern Ocean.

Thomson, chief scientist aboard the *Challenger*, was sceptical of Forbes's azoic theory. He had explored the deep sea before, serving as a naturalist in charge of oceanic research in the Atlantic Ocean and the Mediterranean Sea between 1868 and 1870, aboard the Royal Navy's *Lightning* and *Porcupine*. In his subsequent account of those voyages, he had remarked on how little was known about the depths of the oceans. He was perplexed that, while exhaustive efforts had been made to expand human knowledge of terrestrial environments around the globe, no such effort had been made to understand the 'great ocean slumbering beneath the moon'. He could only surmise that, with each haul taking up to eight hours and requiring absolute attention to prevent too much strain on the dredging rope or engine, the process of dredging in deep water was so laborious as to discourage it. He wrote that the popular perception at the time was that at a certain depth the conditions became

> so peculiar, so entirely different from those of any portion of the earth to which we have access, as to preclude any other idea than that of a waste of utter darkness, subjected to such stupendous pressure as to make life of any kind impossible, and to throw insuperable difficulties in the way of any attempt at investigation.

Nevertheless, in 1869 he had managed to lower the dredge to 4453 metres, and he was confident that the deep ocean would ultimately reveal not a barren waste but an environment inhabited by a rich and varied assortment of marine creatures.[10]

At the time of the *Challenger* expedition the scientific world was still reeling from the revelations of Charles Darwin's *On the Origin of Species*, published in 1859. One of the central questions confronting the *Challenger* scientists was whether the biology of the deep ocean would throw light on some of the questions raised by Darwin's work. A prevailing belief was that, unlike terrestrial environments, the ocean floor was an ancient and unchanging wasteland, and any animals existing there would evolve much more slowly. Darwin theorised that organisms found on land only as fossils might still exist in the deep ocean, and that such 'living fossils' would demonstrate the process of evolution. He had formed his ideas after reading the Scottish geologist Charles Lyell's *Principles of Geology* (the first volume of which had been presented to him by Robert FitzRoy just before the departure of the *Beagle* in 1832), in which Lyell argued that the forces moulding Earth occurred through countless small changes over protracted periods of time, and that those processes were still observable. As the British earth scientist Richard Corfield later observed, the deep ocean at that point became the 'supreme testing ground' of Darwin's theory.[11]

It was a slow voyage. As the *Challenger* tracked back and forth across the ocean basins, daily tasks included recording the sea temperature and salinity, and current flow. In addition, the scientists aboard would drop anchor every 320 kilometres or so to perform magnetic observations; dredging; flora, fauna and geological sample collections; and information gathering about meteorology and the sea floor. In all, they collected examples of about 4417 species from these stations, many of which had never been seen before. Those specimens are held today at the Natural History Museum, in London, where they are still studied by marine biologists.[12] For Thomson, the most 'prominent and remarkable biological result' of the voyage was to prove conclusively that there was no depth limit to the distribution of marine organisms.[13] Animal life existed at all depths and in all places of the world's oceans including the abyssal plain, large flat areas of sea floor at a depth of between 3000 and 6000 metres.

South of the Cape of Good Hope, the *Challenger* voyaged into the cold polar waters beyond the Subtropical Convergence, a region of ocean about latitude 40° South where warm subtropical surface currents meet and mix with the colder subantarctic waters. Remote, cold and stormy, with vast areas perennially covered by ice, the Southern Ocean was always going to be the most challenging of the oceans for scientific voyaging. Indeed, there had been a distinct lack of international interest in investigating the southern polar region since the return of Ross's expedition in 1843. Apart from the sheer logistics of such voyages, in the high southern latitudes there seemed to be little prospect of either suitable land for colonisation or commercial opportunities that warranted government involvement. Matthew Fontaine Maury, the director of the US Naval Observatory and Hydrographic Office, had been one of the few public figures to promote further research in the region, and his focus was on the influence of the Southern Ocean on the weather of the Southern Hemisphere.[14]

The next leg of the *Challenger*'s voyage took the expedition to the most southerly oceans of the globe. They visited the isolated, bleak subantarctic islands and reached as far as the Antarctic continent itself, although Nares was under strict orders to go no further than the edge of the Great Ice Barrier. The expedition's mission was resolutely a quest for scientific knowledge, not territorial expansion, and the hull of the *Challenger* was not designed for breaking through sea ice. Nevertheless, there was much to occupy the scientists. The nature of the deep sea in that region was almost unknown, and in many ways it was the most rewarding part of the voyage. Herbert Swire, a sublieutenant on the *Challenger*, recorded in his journal,

> What with the sea and its denizens, the huge icebergs, the interminable pack, the marvellous sky and the beautiful birds which fly in it my mind is fairly bewildered. Truly they that go down to the sea in ships see Thy wonders in the deep. And yet we are only a few days' steam from Australia and civilization.[15]

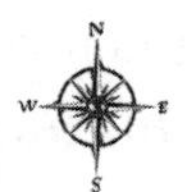

In the polar waters the tow-nets and dredges continued to bring up surprises from the abyss. For Thomson, the abyssal fauna was quite unlike that of adjacent continents, both distinctive and 'singularly uniform' in nature, indicating that it had descended, with subtle variations according to its particular environment, from earlier forms of fauna in the same deep sea. 'The discovery of the abyssal fauna ... seems to have given us an opportunity of studying a fauna of extreme antiquity, which has arrived at its present condition by a slow process of evolution from which all causes of rapid change have been eliminated.'[16]

At around latitude 43° South the *Challenger* tow-nets delivered fine red clay sediments from the ocean floor. The scientists found that they were studded with manganese nodules, thousands of shark's teeth, and the ear bones of whales. Swire noted in his journal that manganese nodules were found in many parts of the world's oceans and that manganese had acquired considerable commercial value because of its use in producing oxygen.[17] Nearing Kerguelen Island, the sediments changed to a pinkish-white ooze made from the shells of strange, tiny foraminifera organisms which inhabited the surface waters; when they died, their shells drifted down like snowflakes into the abyss.[18] Somewhere between the Kerguelens and Heard Island the sediments suddenly changed. John Murray, the Scottish-Canadian naturalist in charge of the collection of biological specimens aboard the *Challenger*, observed that there 'the tow-nets were filled to the brim with a yellow-brown slimy mass, with a distressing odour, through which various crustaceans, annelids, and other animals wriggled'.[19] He recognised the tell-tale yellow ooze of microscopic diatoms, amongst the most common types of phytoplankton (microscopic plants, the basic producers of the Southern Ocean), which had first been identified in freshwater samples in 1783 by the Danish naturalist Otto Friedrich Müller. Equipped with an early microscope to study their form, Müller used extremely fine pieces of silk to filter the diatoms out from the myriad of organisms and plant life he had netted. Diatoms are ancient single-cell plants with cases made of silica. They thrive in cold polar waters, growing beneath icefloes and inhabiting the tiny brine channels that form in sea ice as the surface waters freeze. After death their tiny, delicate cases settle onto the sea floor, and these are found in abundance in the Southern Ocean. They are the primary source of food for larger species, and they underpin the whole of the ocean's ecosystem.[20]

The significance of this diatom ooze dredged from the depths had first come to light in the wake of James Clark Ross's four-year voyage of exploration to the Antarctic. While Ross's voyage is chiefly remembered for his achievement in confirming the existence of the Antarctic continent and charting much of its coastline, it also served to establish the reputation of Joseph Hooker as a leading botanist of his day. Hooker had been just 22 years old when he signed up to Ross's expedition to serve as an assistant surgeon on board the *Erebus*. Ships' surgeons commonly assumed the role of naturalist on long-distance voyages, and Hooker took to the task with relish. At each port of call he collected botanical, zoological and geological specimens, but as the ships drew closer to Antarctica and with fewer opportunities to observe plant life, he began focusing on collecting and drawing specimens of marine plankton. He had just finished reading proofs of *On the Origin of Species*, by his friend Charles Darwin, and was interested in determining the causes of discolouration in sea water. In his journal he described how he examined brown, discoloured ice collected in the Southern Ocean, noting that it released fine sediment when dissolved in water. The sediment comprised 'numerous circular discs [with] opaque centres', which he took to be salt from volcanic dust but later noted that he recognised as diatoms. His equipment on board was not up to the task of a more detailed examination so, after the voyage, he submitted some samples collected between Cape Horn and the Ross Sea for examination by the renowned German botanist Christian Gottfried Ehrenberg, who had published the first paper on diatoms from the waters of the Arctic Ocean in 1838. Thanks to a higher quality microscope,

Ehrenberg confirmed that the yellow stain was indeed a mass of diatoms.[21]

Hooker's book of drawings *Flora Antarctica* was published between 1844 and 1860, in three volumes, and included his observations about the role that winds and currents around Antarctica played in the distribution of these organisms that floated on the ocean currents. He wrote that diatoms were so abundant in the Southern Ocean that they stained the icebergs and sea ice a 'pale ochreous color' wherever they were washed by the swell of the sea.

> The universal existence of such an invisible vegetation as that of the Antarctic Ocean, is a truly wonderful fact … I now class the Diatomaceae with plants, probably maintaining in the South Polar Ocean that balance between the animal and vegetable kingdoms, which prevails on the surface of our globe … The end these plants serve in the great scheme of nature is apparent, on inspecting the stomachs of many sea-animals, as above stated. Owing to the indestructible nature of their shields, they tell their own tale.[22]

In subsequent writings about the profusion of animal and plant life that the *Challenger* scientists dredged up from the Southern Ocean's depths, John Murray claimed the diatoms to be the most interesting. They were the primary source of food in this ocean, and the foundation of all its marine life.[23] The phytoplankton of the Southern Ocean has continued to attract scientific interest, especially since the 1960s, but the sheer vastness of the ocean has made it particularly difficult to determine the extent of its geographical distribution.

Ernst Haeckel, the renowned German zoologist, prepared three of the final reports detailing the *Challenger*'s findings. His radiolaria collection is preserved in the Natural History Museum, London, but perhaps his most enduring legacy is a series of extraordinarily intricate drawings of diatoms and other tiny marine organisms from the deep, many of which he identified himself.[24] The beautiful geometric shapes of diatoms from the depths of the Southern Ocean also found their way into the private drawing rooms of European society as parts of cabinets of curiosities, or *Wunderkammer*, used for exhibiting unusual objects or specimens collected from the far reaches of the globe, and often arranged to tell a story about the natural world. Collectors would provide small microscopes for viewing the delicate forms of diatoms and radiolaria painstakingly arranged with other minute objects, such as butterfly and insect wing scales, and fixed onto exhibition microscope slides. Some slides, such as those prepared by optical instrument makers W Watson & Sons, of London, are still admired by collectors.[25]

Investigating the 'deep bosom of the ocean' was a daunting prospect, as the scientific leader of several deep-sea expeditions, John Gwyn Jeffreys, remarked in 1875. Despite the substantial amount of dredging carried out in various oceans and coastal seas by naturalists on the ships *Lightning*, *Porcupine*, *Shearwater*, *Challenger* and *Valorous*, and the scientific voyages of other nations (including 17 from Sweden alone), Jeffreys observed dryly, 'all that [had] hitherto been effected [had] been to scrape in an imperfect manner the surface of a few scores of acres'.[26] Nevertheless, after Charles Wyville Thomson's death, in 1882, John Murray continued to nurture the legacy of the *Challenger* expedition as a pioneering voyage of oceanography. He oversaw the analysis and publication of the vast quantity of samples and data from the *Challenger* – reports extending to 50 volumes and 600 cases of specimens – and other oceanographic expeditions, and became a forceful advocate for continuing exploration of the deep ocean. When he canvassed the state of knowledge of the world's oceans just before the dawn of the new century, he considered that the deep ocean was becoming the setting for the greatest advances in scientific knowledge of Earth since the voyages of geographical discovery in the fifteenth and early sixteenth centuries.[27]

The seabed had long been imagined as flat and featureless, uniformly covered with fine-grained continental sediments washed into the sea. But by drawing on the measurements and samples gathered during the *Challenger* expedition, Murray was able to construct a rudimentary bathymetric chart revealing a landscape at the bottom of the deep ocean with contours not unlike those of the surrounding continents.[28] It appeared that the ocean floor featured mountains and plains, as well as extensive valleys whose depths, which Murray christened the 'deeps', exceeded 5000 metres. With almost all the marine samples collected over the previous 30 years having passed through his hands, Murray drew on his reputation as one of the most authoritative ocean scientists in Britain to appeal for private sponsorship to support a scheme to explore the little-known depths of the Southern Ocean. In launching his campaign for 'the renewal of Antarctic exploration', in an address in 1894, he argued that it should not simply be to penetrate the icy region with one naturalist on board, in order to make a 'dash at the South Pole'. Rather, it should be 'a steady, continuous, laborious, and systematic exploration of the whole southern region with all the appliances of the modern investigator'.[29]

Others argued for international collaboration in undertaking scientific investigations in the extreme conditions of the high southern latitudes. Karl Weyprecht, who co-led the Austro-Hungarian North Pole Expedition, in 1872–4, believed the poles provided the 'key to many secrets of Nature',

> but as long as Polar Expeditions are looked on merely as a sort of international steeple-chase, which is primarily to confer honour upon this flag or the other, and their main object is to exceed by a few miles the latitude reached by a predecessor, these mysteries will remain unsolved.[30]

Murray's argument for privileging systematic scientific research over 'rapid invasion' was only partly successful, and tensions simmered over the relative importance of scientific research and geographical exploration. Murray continued to advocate for deep-ocean science until his death in 1914, but it was the promise of national glory on the ice-covered Antarctic continent rather than the science of the Southern Ocean that came to define the 'heroic age' of polar exploration between the end of the nineteenth century and World War I.[31]

While this era of Antarctic exploration had a decidedly European bias, Japan also joined in, becoming the first non-European nation to be represented in Antarctica when a privately funded Japanese Antarctic Expedition led by professional army officer Nobu Shirase left Tokyo aboard the *Kainan Maru* (opener-up of the south), in December 1910. The vessel was a small three masted wooden fisherman's sailboat reinforced with steel and supplemented with an auxiliary steam engine, but Shirase managed to reach latitude 74° 16′ South before turning north for Sydney after his passage was blocked by ice in the Ross Sea. He returned to the Antarctic continent in November 1911 aboard the *Kainan Maru* with the aim of undertaking scientific research, but, as the vessel sailed into the Bay of Whales, in the Ross Ice Shelf, Shirase discovered the *Fram* and its crew waiting there for the return of Amundsen's Norwegian expedition party from the South Pole. Shirase decided to take a small team and attempt to reach the pole in January 1912, but he turned back at latitude 80° 05′ South after enduring a week of blizzard conditions. He did, however, plant a Japanese flag in the Ross Ice Shelf, and the east coast became known as the Shirase Coast. He finally met Amundsen in person in the late 1920s although few outside Japan knew of the remarkable expedition.[32]

Despite this nationalistic approach to exploration, the formation in 1902 of the International Council for the Exploration of the Sea signalled the intention of its founding members (Sweden, Norway, Denmark, Finland, Russia, Germany and the United Kingdom) to promote the kinds of sustained investigations of the ocean that the *Challenger* expedition had pioneered. The British biologist Gordon Fogg later observed that the era was defined by a 'new spirit' of inquiry into the marine environment, although he also noted that it was slow to find its way into the Southern Ocean.[33] Most oceanographic research in the region at the time was undertaken largely as a by-product of whaling and Antarctic exploration, and most knowledge was acquired through the time-honoured traditions of dredging, trawling, sampling and measuring as vessels tracked southward into the polar waters. Indeed, trawling remains the main method used for gathering information about life in the Southern Ocean's depths.[34] But plotting ocean currents and dredging animal life were one thing. Directly experiencing the ocean depths was another thing entirely.

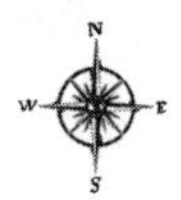

During the 1920s an American zoologist, William Beebe, joined forces with engineer Otis Barton to design a spherical steel ball, called a bathysphere, capable of carrying two occupants into the ocean depths while it remained tethered to a boat at the surface. During the first successful dive in 1930, the bathysphere reached a depth of 243 metres. Beebe later recalled, in the words of Herbert Spencer, feeling like 'an infinitesimal atom floating in illimitable space'. They reached 434 metres in another attempt, the first living humans to enter the dark world of the 'twilight zone'. Beebe pressed his face against the porthole and peered down into the 'black pit-mouth of hell itself'.[35] By 1949 Barton had created a benthoscope, in which he descended to a record depth of 1372 metres in the Pacific Ocean.[36] In 1960, the bathyscaphe *Trieste* reached the bottom of the Challenger Deep, almost 11 000 metres below the surface of the Pacific Ocean at the southern end of the Mariana Trench. It remains the deepest known point of the Earth's oceans.

Oceanographic research halted during World War II, but new technologies that were developed during the war eventually gave fresh impetus to the study of the deep ocean. Maurice 'Doc' Ewing and Joe Worzel, scientists employed during World War II at the Woods Hole Oceanographic Institution in the United States, developed a continuous echo sounder for use by the US Navy to track the movement of enemy submarines. This technology revolutionised the study of the deep ocean, as it enabled scientists to record continuous depth measurements in deep water by aiming an electronic ping at the seabed and determining the time required for the sound waves to travel from the surface to the bottom and back. The trajectory of the ping and the returning echo were recorded by a stylus emitting an electric spark onto a spooled strip of paper. Ewing oversaw many thousands of such measurements made across the North Atlantic in the years immediately after the war.

The 1950s proved to be a revolutionary time in oceanography, during which popular perceptions of the deep ocean changed. In 1950 the US Mid-Pacific (or Mid-Pac) Expedition, led by Roger Revelle, explored the topography of the ocean bottom. The results confirmed that the sea floor was not only mountainous but also relatively youthful, compared with the antiquity of the continental landmasses.[37] In the same year, Denmark launched a national scientific expedition to investigate the extreme depths of the world's oceans. As Anton Bruun, the leader of the 1950–2 *Galathea* expedition (the second of three expeditions known by this name) put it, the ocean depths were the only 'white spots' left on the world map. The expedition intended to record sea temperatures and currents and to collect data on plant life. Perhaps anticipating that his audience might not be excited by this, Bruun claimed that it would also investigate whether there was any truth in the 'old belief' in sea monsters or sea serpents. 'Just think what might be down there', he said, 'unknown and unseen by human beings since the beginning of time itself.'[38] It seems that Darwin's image of the deep ocean as a repository of 'living fossils' was still compelling nearly a century later.

The *Challenger* expedition had shown that the ocean depths contained a rich and varied sea life, both in the deeper water and on the seabed, at least as far down as 5000 or 6000 metres. By the mid-twentieth century ocean scientists had developed a reasonably detailed understanding of the marine life in depths of between 1000 and 4000 metres, but they still knew very little about whether living organisms could survive in water below 4000 metres, and 'virtually nothing' of marine creatures that could survive in depths of more than 6000 metres. For its deep-sea mission, the vessel *Galathea* was fitted with an exceptionally long, 22-tonne steel wire rope which, once paid out, took up to three days to wind back in. There was also an assortment of nets, trawls, grabs and other instruments to 'sweep the sea bottom to a depth greater than the height of Mt. Everest'. The reference was timely, as the first successful ascent of the highest mountain in the world was made in 1953, a year after the return of the *Galathea* expedition. There was an elegant symmetry to it: at the same time, one expedition was aiming to reach the deepest part of Earth, and another to reach the highest. Australian newspapers in country towns from Burnie to Broken Hill eagerly reported that the Danish expedition would spend two years exploring the depths of nearly every sea and ocean of the world in order to 'clear up once and for all' whether or not sea monsters existed.[39]

Sea monsters and serpents have long captured the Western popular imagination and attracted the interest of natural scientists and writers alike. One of the most renowned novels of the nineteenth century was the French writer Jules Verne's *Vingt mille lieues sous les mers: tour du monde sous-marin* (*Twenty Thousand Leagues Under the Sea: A Tour of the Underwater World*), serialised in 1869–70 and published in 1871, in which an expedition sets out from the east coast of the United States and into the Pacific Ocean around Cape Horn in search of a mysterious sea monster which, they soon discover, is in fact a submarine. Three of the expeditioners are captured, and they are taken on a voyage through the depths of the world's oceans as far as Antarctica, doing battle with a school of giant squids along the way.[40] In 1892 Anthonie Cornelis Oudemans, director of the Royal Zoological and Botanical Gardens at The Hague, published a more scientific survey of 187 reported sightings of sea serpents in the world's oceans between 1555 and 1890, and dedicated it to the 'owners of ships and yachts, sea captains and zoologists'.[41] Sea monsters were commonly reported as being seen in both hemispheres, and such stories never failed to attract public attention.

In Australia, reports of sightings often came from as far away as Scotland. Occasionally they occurred much closer to home, and the remote and sparsely settled coastlines and estuaries of southern Australia were favourite locations. In 1913 two men encountered a strange animal as they walked along a Southern Ocean beach on Tasmania's remote western coast. They described the creature as having chestnut-coloured fur and measuring about 4.5 metres long. It had the head of a dog and a thick, arched neck, although it 'bore no resemblance to seals or sea leopards'. As they approached, it reared up on its hind legs and bounded into the sea. It would have come as no surprise, then, when the front page of the Launceston *Examiner* declared, 'Scientists to hunt sea monsters'. Mythical monsters from the deep had suddenly become entangled with scientific investigations of the undersea world.[42]

The *Galathea* spent 21 months trawling the world's oceans, collecting data and specimens from as deep as 10 000 metres. Einer Steemann Nielsen, who pioneered the use of carbon-14 (^{14}C)

to measure the productivity of oceanic waters in different regions, found high fertility in the Southern Ocean off South Africa and concluded that the ocean was likely to have a similar productive capacity to that of the land.[43] Bruun's eloquent account of the voyage, published as 'Animal life of the deep sea bottom', revealed that samples of mud from the perpetual darkness of the abyssal plain that were brought to the surface by the Petersen grab (a mechanical device used to obtain samples from the bottom of a body of water) contained bacterial flora. The hauls may not have produced 'sea serpents', but they revealed whole communities of animal life living in the greatest ocean depths.[44]

By the mid-twentieth century ocean scientists, reliant on technology and a good dose of fortitude, were opening a window onto the deep Southern Ocean, transforming the view from that of a two-dimensional surface of winds and fog and ice to one of a dynamic three-dimensional undersea world teeming with life. Enthusiasm for deep-ocean exploration spilled over into public life, even in Australia where most settler Australians still identified with the heroic narratives of colonising the outback. The Western Australian writer Tim Winton noted that the people of this island continent, surrounded by three of the great ocean basins of the world and living for the most part around its edges, had never really been ocean people.[45] Nevertheless, the scientific discoveries of the 1950s seemed to stir public enthusiasm, and there were others – adventurers, explorers and fishers – who were drawn to explore the deep ocean for themselves.

During World War II Frenchmen Jacques-Yves Cousteau, a naval officer, and Emile Gagnan, an engineer, developed the aqua-lung, giving divers the freedom to reach greater depths than humans had ever ventured before.[46] Cousteau went on to become a prominent advocate for ocean conservation, evoking images of the sea and people's deep relationship with it. 'Our flesh is composed of myriads of cells,' he wrote in 1976, 'each one of which contains a miniature ocean … comprising all the salts of the sea, probably the built-in heritage of our distant ancestry.'[47]

A number of popular books about ocean exploration published around this time were written by scientists aiming to give general readers a taste of the undersea world.[48] Some, written with a lyricism that has all but disappeared from modern scientific prose, became best-sellers. In *Science of the Seven Seas* published in 1945, the renowned American oceanographer Henry Stommel wrote lovingly of the slimy oozes to be found in the deep sea. Under the heading 'Oodles o' ooze', he described the myriad of tiny floating organisms 'who, when alive, frolic and swim through all the seven seas, and who, when overtaken by old age and death, fall to the ocean bottom to become part of the eternal ooze'. The deep ocean was a 'kind of junk heap' of time, where an oceanographer might discover 'the tooth of a prehistoric shark' alongside 'cinders dropped from a steamer', the remains of a meteor from 'outer space', and rocks carried in an iceberg 'from a distant continent'.[49]

According to his colleagues, Stommel had an uncanny ability to see the ocean in three dimensions. In 1957 he sailed into the Sargasso Sea near Bermuda with John Swallow, who had devised a method of measuring currents at all depths using floats that emitted pings that could be tracked using underwater microphones and detected from a nearby ship. As Swallow deployed the floats, he watched them scatter in all directions, observing that there was something interesting going on in the depths.[50] In a subsequent experiment in 1973 Swallow's floats showed that, far from being a dark and silent world, the deep ocean was a chaotic environment, regularly convulsed by violent eddies and abyssal storms that could rage for days. Just as Matthew Fontaine Maury had predicted more than a century earlier, the ocean had its own weather.[51]

The American marine biologist Rachel Carson is perhaps best known for her book *Silent Spring*, in which she drew attention to the effects of pesticides on the environment. She also published three books about the oceanic environment. In the second and most popular book of the series, *The Sea Around Us*, she began with the gentle observation that humans, unable to return to their own marine origins, could 're-enter [the ocean] mentally and imaginatively' by drawing on their own skill and ingenuity, accompanied by a sense of wonder.[52] Carson's gifts as both writer and scientist enabled her to reach a wide audience with a heady mix of fact and emotion. By interweaving science and storytelling, she sought to recast the simple animal and plant life of the sea as a complex, fragile and beautiful world that demanded humans' understanding and protection.[53] Her books focused on the oceans of the Northern Hemisphere, although she briefly ventured into the Southern Ocean to describe how the Roaring Forties scoured the surface while, deep below, the cold bottom current moved 'ponderously slow, the measured creep of icy, heavy water', on its 'global wanderings' across the world's oceans. 'Perhaps in these sunless streams the weird inhabitants of deep waters drift, generation after generation, surviving and multiplying because of the almost changeless character of these slowly moving currents.'[54]

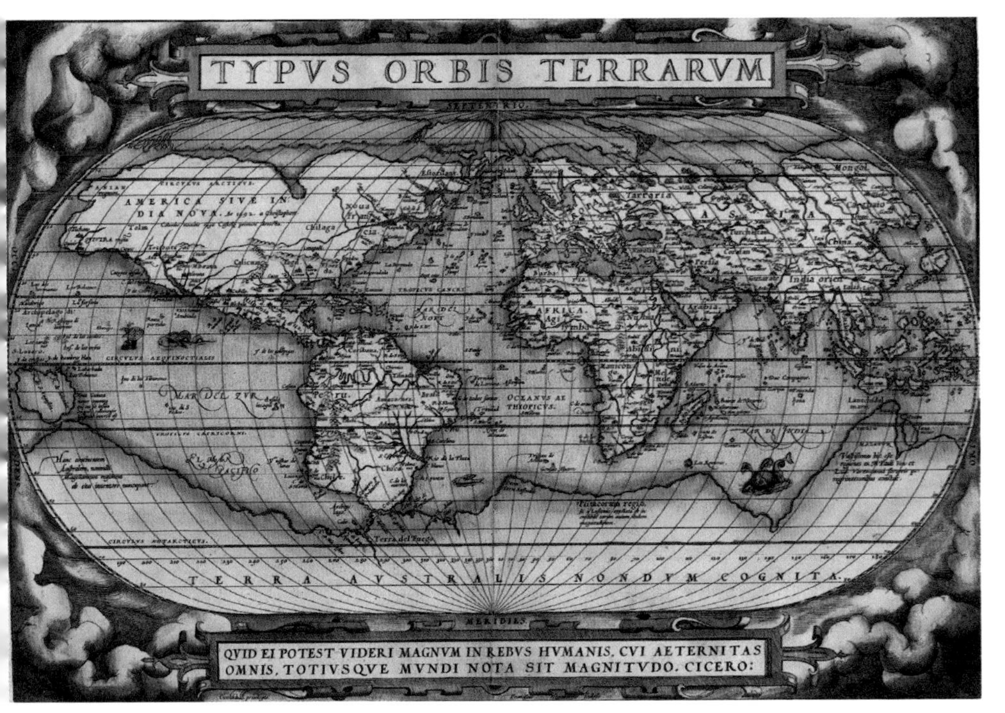

1. This map of the world was produced by Abraham Ortelius as part of a series published in *Theatrum Orbis Terrarum* (Theatre of the World) in 1570. It depicted 'Terra Australis nondum cognita' (Unknown South Land) as a landmass extending from the South Pole to the Tropic of Capricorn, reflecting European ideas about the geographical nature of the world at that time.

Library of Congress, Control No. 98687183

2. This illustration, from Captain James Cook's *A Voyage Towards the South Pole* published in 1777, shows the *Resolution* at anchor beside an 'ice island' in the Southern Ocean as crew members shoot at seabirds for food and break up ice for drinking water. It was drawn from nature in 1773 by the artist William Hodges, who accompanied Cook on his second voyage of exploration between 1772 and 1775.

National Library of Australia Pictures Collection, nla.obj-135699564

REGION FROIDE:
CANADA
MER DE FRANCE:
MER DESPAIGNE:
MER OCCEANE:
DE CANCER
MER DES ENTILLES:
EVROPPE
LE PERV:
LA LIGNE:
LA ZONA TOR RIDA:
MER DV SV:
AMERIQVE
TROPIQVI
DE:
MER AVSTRALLE:
MER DE MAGELLAN:
LA TERRE:

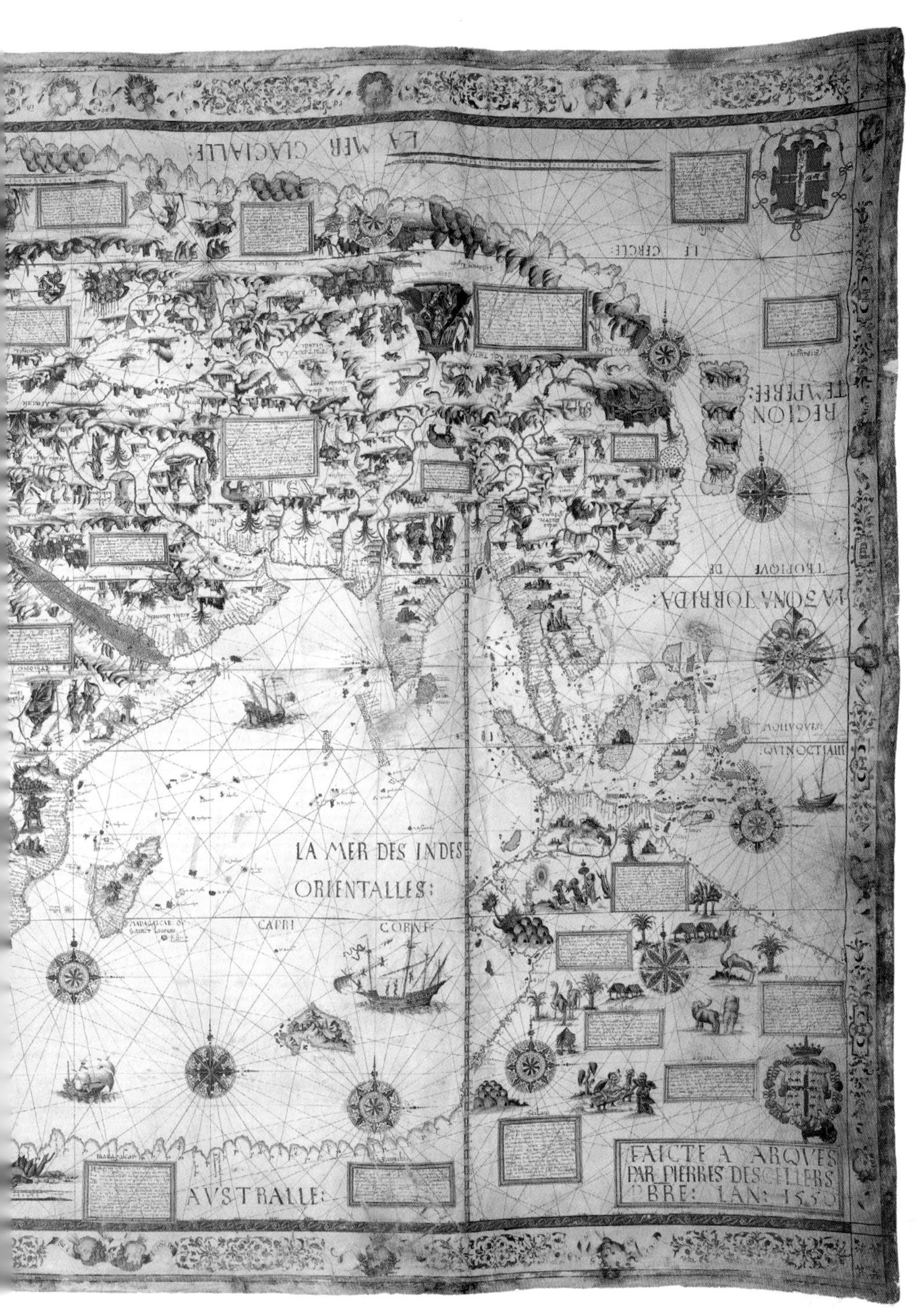

3. Pierre Desceliers produced this lavishly illustrated map in 1550 as a work of art rather than for navigational purposes. He combined observed and mythical elements to portray perceptions of the physical nature of the world beyond Europe in the sixteenth century. In the region of the as yet unexplored Southern Ocean, he depicted trading vessels and sea monsters near the coast of La Terre Australle (the southern land), an imaginary landscape inhabited by wild creatures.

British Library, Add MS 24065, BL3703834

4. Each year one of the largest colonies of king penguins (*Aptenodytes patagonicus*) on Earth gathers at St Andrews Bay, South Georgia to breed and raise chicks. Some scientists warn that these glorious creatures of the Southern Ocean are likely to face extinction as a result of global warming and the depletion of Antarctic krill, their main food source.

Photograph by Joy McCann

5. Dr William Ingram, medical officer and biologist with two British, Australian and New Zealand Antarctic Research Expeditions conducted during 1929–31, with a young wandering albatross on the Crozet Islands. The steep, rugged island group, 2400 kilometres from Antarctica, is part of the French Southern and Antarctic Territories and has been designated a national conservation area.

National Library of Australia, nla.obj-141179428

6. TOP LEFT Conrad Martens painted this scene of local inhabitants near the *Beagle* in the channels of Tierra del Fuego in 1831. Martens spent 15 months as official artist with the second *Beagle* voyage while on his way to Australia. The voyage is best known for its part in British naturalist Charles Darwin's scientific theory of evolution by natural selection.

Cambridge Digital Library

7. LEFT A sealer removes the blubber from a young southern elephant seal (*Mirounga leonina*) on Tristan da Cunha, circa 1824. Southern elephant seals spend most of their time cruising around the Southern Ocean, visiting subantarctic island beaches twice a year to breed and moult. They are powerful swimmers, capable of diving to depths of up to two kilometres. They are cumbersome on land, however, which made them easy prey for commercial sealing gangs in the nineteenth century.

National Library of Australia, nla.obj-134494616-1

8. ABOVE Dr Edward Wilson painted the Great Ice Barrier (Ross Ice Shelf) in January 1911 during the *Terra Nova* expedition led by Robert Falcon Scott. A qualified doctor, Wilson was one of five men to reach the South Pole in January 1912. He died on the Barrier alongside Scott and Lieutenant Henry 'Birdie' Bowers on their return journey, leaving an extraordinary legacy of paintings and drawings from the 'heroic' era of Antarctic exploration.

National Oceanic and Atmospheric Administration collection

9. Dr Edward Atkinson in his laboratory at Cape Evans, McMurdo Sound on 15 September 1911 during the *Terra Nova* expedition led by Robert Falcon Scott. One of the expedition's aims was to expand scientific knowledge about the region.

Alexander Turnbull Library, Wellington, New Zealand, PA1-f-067-046-2

10. Icebergs come in a variety of shapes and colours in Cierva Cove near Graham Land on the Antarctic Peninsula.

Photograph by Joy McCann

11. Frontispiece from Jules Verne's *Vingt mille lieues sous les mers: tour du monde souse-marin* (*Twenty Thousand Leagues Under the Sea: A Tour of the Underwater World*), by artist J. Hetzel, Paris, 1871.

Courtesy of Harvard University

12. LEFT Naturalist reading deep-sea thermometers during the voyage of the HMS *Challenger* (1872–6) to determine the temperature on the floor of the world's oceans.

National Oceanic and Atmospheric Administration

13. BELOW This image taken by the Australian photographer Frank Hurley in 1930 shows scientists inspecting a haul dredged from the Southern Ocean on board *Discovery* during the first British, Australian and New Zealand Antarctic Research Expedition in 1929–30. From left: Sir Douglas Mawson (expedition leader), Dr William Ingram (medical officer and biologist), an unidentified man, and James Marr (oceanographer).

Australian Antarctic Division

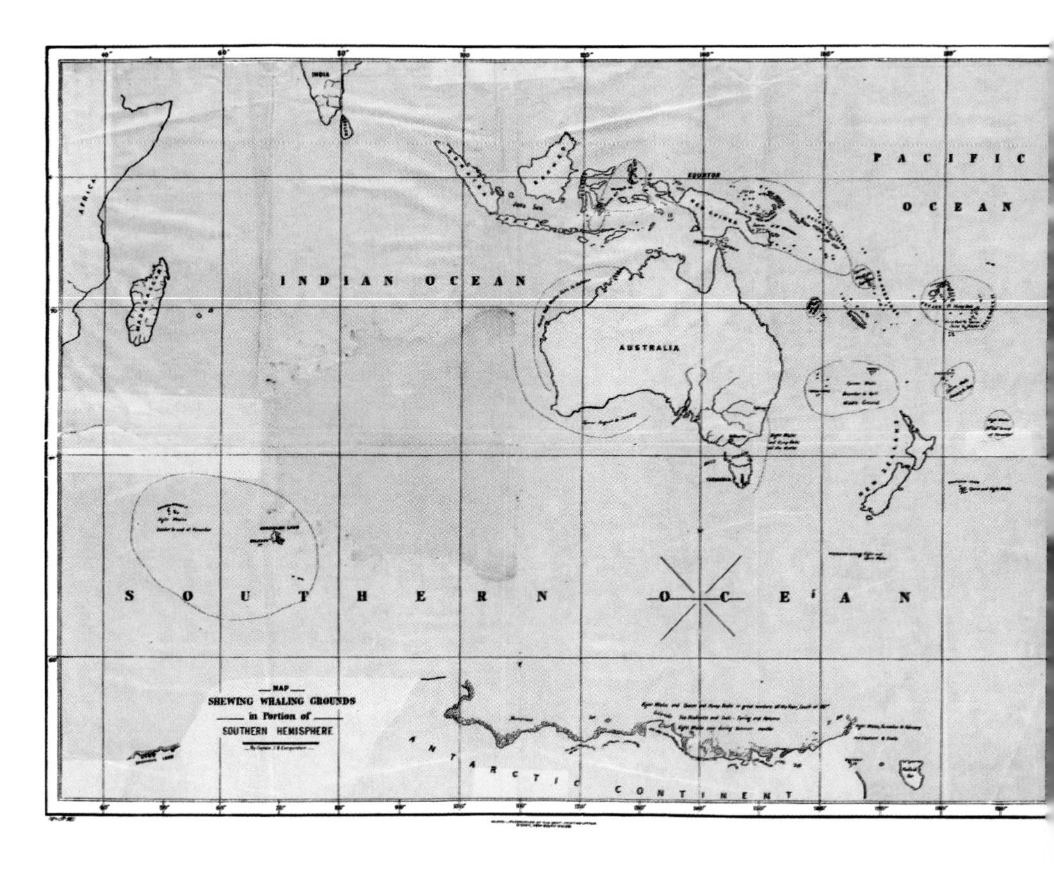

14. ABOVE Map of known whaling grounds in the Southern Hemisphere in 1892 by JB Carpenter, captain of the whaling vessel *Costa Rica Packet* which operated out of Sydney between 1888 and 1891. The map shows seasonal locations of sperm, humpback and southern right whales, fur seals and elephant seals as the southern whaling industry was expanding.

National Library of Australia, nla.obj-23131932

15. OPPOSITE ABOVE *Natives of Encounter Bay making cord for fishing nets, in a hut formed of the ribs of a whale, 1847*. The British artist and naturalist, George French Angas, produced a series of paintings depicting people and the natural environment during his travels in the coastal regions of South Australia. This painting was reproduced in the 1847 edition of *South Australia Illustrated*, part of Plate no. 56 (lithograph by JW Giles).

National Library of Australia, nla.an7350676

16. OPPOSITE BELOW William Duke's painting shows whaling in the Derwent River off Hobart in 1849. In 1804 observers reported so many whales in the river that it was too dangerous for boats to navigate. By the 1840s, the Tasmanian whaling industry was in decline due to over-hunting.

Tasmanian Museum and Art Gallery collection, AG2997

NATIVES OF ENCOUNTER BAY.

MAKING CORD FOR FISHING NETS, IN A HUT FORMED OF THE RIBS OF A WHALE

17. Whalers stand beside a blue whale (*Balaenoptera musculus*) on the flensing plan at Grytviken on South Georgia. This 1914 photograph was taken by Frank Hurley during the Imperial Trans-Antarctic Expedition in 1914–17 led by Ernest Shackleton when blue whales were still abundant in the Southern Ocean. An estimated 300 000 blue whales were killed in southern oceans for their oil-rich blubber between 1900 and the 1960s.

National Library of Australia, nla.obj-158917162

18. Antarctic krill (*Euphasia superba*) are a 'keystone' species of the Southern Ocean's ecosystem, and the principle source of food for whales, seals, penguins, squid and fish. They breed in such large numbers that their total weight (or biomass) is more than that of the entire human population of the planet.

Photo by David Tipling

19. Local children welcoming international delegates to the meeting of the Commission for the Conservation of Antarctic Marine Living Resources held in Hobart, Tasmania in October 2016. The sign reads: 'Dear CCAMLR delegates, We support marine protected areas in the Southern Ocean. Yours faithfully, Over 1 million global citizens. P.S. Please don't delay.'

Reproduced by kind permission of Rob Blakers

When *The Sea Around Us* reached Australia's shores, the *Sydney Morning Herald* published a review, noting that humans had mapped and catalogued almost every corner of Earth's surface and had even succeeded in exploring the ocean's 'twilight zone', as when Otis Barton descended to 1372 metres in his benthoscope in 1949. Yet, 'beneath the oceans, there [was] an immense territory into which scientists [were] only now beginning to probe'.

> From far below this level, nets and scoops have brought up strange and fantastic creatures, and from this evidence scientists have begun to mortice together a theory that horizontal zones or 'communities' of sea life populate the depths, living parasitically on each other and all ultimately dependent on a slow rain of food particles and plant life from above.[55]

The *Australian Women's Weekly* declared that no home library should be without a copy of Carson's book. Sydney's *Daily Mirror* serialised it in 1952, promising readers that they would 'find themselves explorers of the deep, thrilled by a sense of great discovery'.[56] Australia's first underwater filmmaker, Noel Monkman, shot the underwater sequences for the film version of *The Sea Around Us*. Although Carson dissociated herself from the film because of scientific inaccuracies, it was nevertheless the first of many popular films and documentaries to explore the human dimensions of the ocean and to bring the deep sea into people's living rooms.

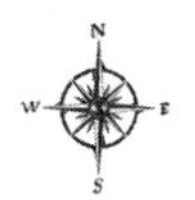

The prospect of a new ocean frontier was whetting the public appetite for stories about deep-sea exploration, but acquiring knowledge about the topography of the ocean floor still relied on the painstaking tasks of taking soundings using hemp or wire lines and lowering dredges and bottom-sampling devices. The traditional method of sounding involved dropping a hemp rope overboard, weighted down with a lead object such as a cannonball. It was a slow and tedious task, and plotting individual soundings along a ship's track could yield only a fragmentary picture of the seabed following a specific line. This is how Matthew Fontaine Maury had conducted a series of 200 depth soundings in the North Atlantic Ocean for the US Navy in the 1850s. When he transferred his measurements onto a chart in 1854, the contours revealed that a plateau rose in the middle of the ocean, indicating that the seabed was perhaps not as flat and featureless as many imagined.[57] Two decades later, the *Challenger* expedition produced a series of depth soundings and sea temperature measurements in the central Atlantic Ocean indicating that a broad mountain ridge ran along the bed, dividing the ocean into two distinct basins. Further south, the Antarctic region also offered some curious geological evidence about the nature of the ocean floor.

When the geologist WT Blanford wrote a chapter for *The Antarctic Manual for the Use of the Expedition of 1901*, produced to guide the scientific investigations of Robert Falcon Scott's *Discovery* expedition, he suggested that they should look for evidence of the *Glossopteris* fern.[58] The plant was known to have grown throughout the southern continents during the Permian period, between 299 and 252 million years ago, and its discovery in Antarctica would provide the crucial piece of evidence to support the theory proposed by Austrian geologist Eduard Suess that land bridges had once connected South America, Africa, India, Australia and Antarctica, forming an ancient continent that he had called Gondwanaland.[59] Scott did discover *Glossopteris* in Antarctica, although it was during the *Terra Nova* expedition in 1910–13, and the significance of the discovery was overshadowed by the tragic events of Scott's return journey from the South Pole in 1912.

The party had stopped at Mount Buckley, on Buckley Island, to spend a few hours geologising amongst the rocks of the Beardmore Glacier. Scott noted in his diary, 'From the last Wilson, with his sharp eyes, has picked several plant impressions, the last a piece of coal with beautifully traced leaves in layers, also some excellently preserved impressions of thick stems showing cellular structure.'[60] When the party sent to rescue Scott and his companions finally located their tent, they discovered the frozen bodies of Scott, Edward Wilson and Henry Bowers, together with a sledge filled with 16 kilograms of rock samples that they had hauled from the glacier. The rocks were subsequently found to contain a fragment of a fossilised feather-shaped leaf of the *Glossopteris*. Similar fossils discovered across the southern continents indicated that they had supported an abundance of flora and fauna until about 55 million years ago, when Australia moved northward, isolating the Antarctic landmass. Remnant forms of vegetation still growing in the southern cold-temperature rainforests of Tasmania, New Zealand and southern South America also existed in Antarctica, at least until about 35 million years ago. Evidence to show when the Antarctic vegetation died out is still fragmentary, and debate continues about just when the environmental conditions of the region changed.[61]

The idea that continents drifted across the face of Earth was controversially proposed by the German geophysicist Alfred

Wegener, in 1912.[62] Wegener theorised, as did Francis Bacon in 1620, that since the coastline of eastern South America seemed to fit so exactly with that of western Africa, the two southern continents were likely to have once been joined. Furthermore, he discovered that geological features on the separate continents were identical. He developed the idea further by suggesting that all of the continents had once formed a single landmass, which he called Pangaea (from the Greek meaning 'all the Earth'), that Pangaea had slowly broken apart about 300 million years ago, and that the fractured landmasses were gradually moving away from each other. The problem with this continental drift theory was that there was no plausible explanation to support it, and the theory so divided the scientific community that even to utter the word 'drift' was to be considered a scientific heresy.[63] Wegener proposed gravitational forces as an explanation, but those opposed to the idea – the 'non-drifters' – argued that it was a physical impossibility for solid rock to move. In 1948 a young geologist and mathematician called Marie Tharp began working at the Lamont Geological Observatory (since renamed the Lamont-Doherty Earth Observatory) at Columbia University, New York. Maurice Ewing, the American geophysicist and oceanographer who had helped develop the continuous echo sounder, recruited Tharp to draft and plot ocean floor profiles for a young oceanographer called Bruce Heezen; they formed a partnership that would revolutionise scientific understanding of Earth's geological history and provide crucial evidence to support the theory of continental drift.

Tharp's father was a soil surveyor working for the US Department of Agriculture, and she had spent her childhood moving around the country with him as he collected data to make soil survey maps. Map-making, she said later, was just in her blood.

Nevertheless, women of her generation had few opportunities to pursue a career in science, and it was only with the United States' entry into the war, after the bombing of Pearl Harbor, that she became able to pursue her interests. The University of Michigan, finding its geology department depleted of men who had signed up for military service, began offering positions to women. Tharp commenced her work compiling a topographical map of the ocean floor using the vast archive of sounding data from numerous voyages. She began by drawing sections of the North Atlantic sea floor, plotted from the spider's web of ships' tracks, and continued with other ocean basins. 'The whole world was spread out before me (or at least, the 70 percent of it covered by oceans)', she recalled.

> I had a blank canvas to fill with extraordinary possibilities, a fascinating jigsaw puzzle to piece together: mapping the world's vast hidden seafloor. It was a once-in-a-lifetime – a once-in-the-history-of-the-world – opportunity for anyone, but especially for a woman in the 1940s. The nature of the times, the state of the science, and events large and small, logical and illogical, combined to make it all happen.

When she compared the six profiles of the North Atlantic Ocean seabed, she noticed an unexpected feature: each showed the oceanic ridge with a V-shaped depression in the centre. 'The individual mountains didn't match up', she recalled, 'but the cleft did, especially in the three northernmost profiles. I thought it might be a rift valley that cut into the ridge at its crest and continued all along its axis.' She decided to produce a world map that revealed the topography of the seabed without its blanket of water. Germany's oceanographic *Meteor* expedition, in 1925–7, had pioneered use of the echo sounder. Drawing on the expedition's results, Tharp was able to show that the Mid-Atlantic Ridge formed part of a global oceanic ridge system, a submarine mountain range 80 000 kilometres long which snaked around the world's oceans about halfway between the continents on either side, and that it had a distinctive valley running along its centre. 'There was but one conclusion: The mountain range with its central valley was more or less a continuous feature across the face of the Earth.'[64]

Inspired by this work Harry Hess, a Princeton University geologist who had served with the US Navy in World War II, and Robert S Dietz, a scientist with the US Coast and Geodetic Survey, proposed that the mid-ocean ridge marked the boundary between tectonic plates. As a result of volcanic activity, they reasoned, molten rock oozed up from beneath Earth's mantle between the plates and formed new crust as the older rock spread away from the ridges and sank into the mantle. Hess's theory of sea floor spreading transformed understanding of the geological history of Earth, providing an answer to several puzzling aspects of the ocean floor.[65]

The National Geographic Society, keen to communicate these discoveries to a wider audience, commissioned Heinrich Berann to draw a panoramic image of the Indian Ocean seabed. Berann, an Austrian artist who specialised in producing alpine panoramas to promote skiing to tourists, used Tharp's drawings as the basis for his artwork. The result was so successful that he collaborated with Tharp and Heezen between 1967 and 1975 to produce a version for each of the major ocean basins. Finally, in 1977, he produced a panoramic map of the entire world ocean showing the mid-oceanic ridge extending around the seabed, bisected by hundreds of fracture zones as though Earth's crust had been torn apart and stitched up again. In the Southern Hemisphere, the mid-oceanic ridge split the ocean floor into several distinct plates showing, for the first time, the ancient bonds between the continents that had once formed Gondwana. As Tharp reflected later, the maps were crucial in enabling both scientists and the general public to visualise a part of Earth that would otherwise have remained hidden.[66]

Bruce Heezen also became involved in a study of the deep ocean with another colleague, Charles Hollister.[67] Their project began in 1947 'out of pure curiosity'. They were keen to reveal the living face of the abyss that, with post-war technological advances, had become an exciting new frontier for ocean science. Heezen and Hollister combined knowledge acquired from previous deep-sea voyages with recent theories about the movement of ocean currents and volcanic activity in Earth's oceanic crust. Their most intriguing source was a series of photographs of the sea floor taken with deep-sea flash cameras illuminating features never before seen in this 'seasonless abyssal world'. Amongst the thousands of photographs they examined were some that revealed the outlines of ripples and scour marks blanketed in the soft sediments that rain down unceasingly, casting a veil over the sea floor. They brought to life creatures tantalisingly small and out of focus in the murky depths, adapted to conditions so precarious that they had long been thought to not sustain life at all. Heezen wrote, 'One's dim memory demands: But where are all the monsters of the deep? Where are the rocky ribs of the earth? Where are those gems of purest ray serene, which the caves of ocean bear? Where are all the bizarre and romantic seascapes?'

After 20 years of studying the growing archive of fuzzy photographs, the scientists made their first descent in a submersible, amazed to see the sea floor come into focus for the first time.

With their eyes glued to the portholes, they could see the strange inscriptions in sediments and vague outlines of creatures – fish, starfish, spiders, clams, snails, worms and crabs, 'the shepherds of abyssal herds' – all feeding on the 'bacterial pastures on the endless refuse heap of the deep-sea floor'. 'At last we could feast our eyes on the oozy bed. We could see if the animals moved, what tracks they made, and could sometimes determine if the burrows were inhabited.'

They chose a selection of the photographs for their book *The Face of the Deep*, published in 1971. Around one-third of these came from an extensive photographic survey of the Southern Ocean floor taken during the US National Science Foundation's research voyage aboard the *Eltanin* between 1962 and 1975, which produced more than 10 000 photographs taken at 500 locations.[68] The deep ocean had long been imagined as a quiet place, relatively unperturbed by the strong surface currents of the West Wind Drift, but the photographs proved otherwise. The images revealed how the violent force of the Antarctic Circumpolar Current left its dramatic signature in the abyssal sediments of the Bellingshausen Basin and Scotia Sea and in the scoured depths of the Drake Passage. They showed that the currents of the Southern Ocean ran deep, sculpting abyssal clay and silt into fluted lines and crescents as they swept around rocks and hardened mud on their global journey to the far corners of the planet, imprinting their indelible tracks across the floors of the world's oceans.

6

CURRENT

For the animal shall not be measured by man. In a world older and more complete than ours they move finished and complete, gifted with extensions of the senses we have lost or never attained, living by voices we shall never hear. They are not brethren, they are not underlings; they are other nations, caught with ourselves in the net of life and time, fellow prisoners of the splendour and travail of the earth.

Henry Beston, *The Outermost House*, 1928[1]

Latitude 54° 38′ South, Longitude 35° 47′ West: Gold Harbour, South Georgia (4 November 2017)

The beaches of South Georgia may be a feast for the senses, but the surrounding ocean is curiously quiet. It is spring, and the whales are still making their way south from warmer waters. I imagine the bays all around filled with the sounds of their blowing. Other sounds once reverberated in these bays, when the whalers were here. We anchor off the island's southeastern coast, adjacent to a series of vertical ice cliffs that reveal the bare teeth of crystal glaciers streaked with blues and greys. A leopard seal grins as it surfaces briefly before slipping silently beneath the boat. On the beach a giant petrel suddenly launches at a king penguin chick and quickly kills it, plunging its beak into the still-warm body. Perhaps the petrel sensed that the chick was ill or weak. Perhaps it is simply too hungry to wait for something to die. Life is brutal here, even without the whalers. Eat or be eaten. The waters may be quiet but underneath they are brimming with life.[2]

The island lies directly in the path of the Antarctic Circumpolar Current, and the mighty current produces an upwelling of deep-ocean water as it meets the underwater mountains of the Scotia Ridge, bringing a rich bounty of nutrients to the surface and creating massive blooms of phytoplankton. The Antarctic Convergence to the north and another front to the south add to the proliferation of marine life in these waters. Swarms of krill thrive here, along with other zooplankton (tiny animals that drift with water currents), in preparation for a feast like no other. Some of the largest concentrations of whales on Earth – humpback, fin, sei, southern right, minke and blue plus several species of toothed whale – have traditionally congregated here for the summer smorgasbord. Some spend the whole season here, while others linger briefly, on their way to the Antarctic ice.[3]

The planetary ocean is like a body with a network of arteries and veins of water, currents invisible to the eye but constantly flowing, carrying large masses of salt water around the world in different directions at varying speeds and depths. Some move horizontally, whipped up by strong winds conspiring with the rotation of the planet to create powerful surface currents. Others are vertical currents, carrying deep bottom water from around Antarctica and the North Atlantic: as the surface polar ocean waters cool or freeze in winter, dense water is formed that sinks to the ocean floor and flows along the bottom, into the world's temperate oceans.[4]

A simple way of picturing the movements of the invisible currents of the Southern Ocean is to imagine taking a voyage southward, counting the lines of latitude from north to south as we go from the warmer equatorial waters to the coldest waters circulating around the glaciated continent. South of latitude 40° South we encounter the circumpolar storm track, the turbulent domain of the Antarctic Circumpolar Current.[5] With no continent to contain its flow, this is the largest ocean current on Earth. It is estimated to be up to 200 kilometres wide and extend from the surface to depths of 4000 metres, carrying enough water to fill all the world's rivers at least 150 times over.[6] In some places major undersea ridges rise and troughs descend along the boundaries of or within tectonic plates, interrupting the circumpolar current's flow and forcing upwelling of deep water around the Macquarie Ridge Complex, the Scotia Arc and the Kerguelen Plateau.[7]

Between the southern tip of Tierra del Fuego and the northern tip of the Antarctic Peninsula, the circumpolar current squeezes through the Drake Passage, a channel 800 kilometres wide. The Passage opened between 49 and 17 million years ago – scientists are uncertain about the timing – as the ancient southern supercontinent of Gondwana fragmented. The deep oceanic passage of the Tasmanian Gateway, between Australia and Antarctica, was also formed.[8] The Southern Ocean was born, and it became a force to be reckoned with.[9] Oceanographers estimate that its giant current surges through the Drake Passage at 103 million cubic metres of water per second.[10]

If we can resist being drawn into the circumpolar current's easterly path, we continue our journey, jostled by its turbulent winds and mountainous waves. Invisible to the human eye, the undersea tracks of the ocean's water masses are complex and compelling. Navigating southward, across the current, we pass through a series of jet streams – some broad and slow, others fast and narrow – separated by bands of slower moving water. The most significant of these are the Subantarctic Front, the Antarctic Polar Front and the Southern Antarctic Circumpolar Current Front.

South of the Antarctic Polar Front lies the distinctive biological region of the Antarctic Convergence, where cold polar waters from the south meet and slide beneath the warmer, less dense, temperate waters from the north.[11] This slow exchange takes place along a narrow, watery frontier that wobbles and moves – northward or southward, depending on the season – like an overextended conga line. We can detect these fronts as we pass from north to south by noticing changes in the temperature, salinity, density, oxygen and nutrients of the different water masses, as well as from the appearance of species of birds, fish and animals that are rarely found north of the convergence.[12]

During his second voyage into the Southern Hemisphere, in December 1772, James Cook recorded a sharp drop in temperature at latitude 48° 41′ South, noting the 'sudden transition from warm, mild weather, to extreme cold and wet [that] made every man in the ship feel its effects'. It also killed most of the sheep, hogs and geese he had taken on board at the Cape of Good Hope to supply the crew with fresh meat for their journey into

Oceanographic features, seas and landmarks in the high southern latitudes

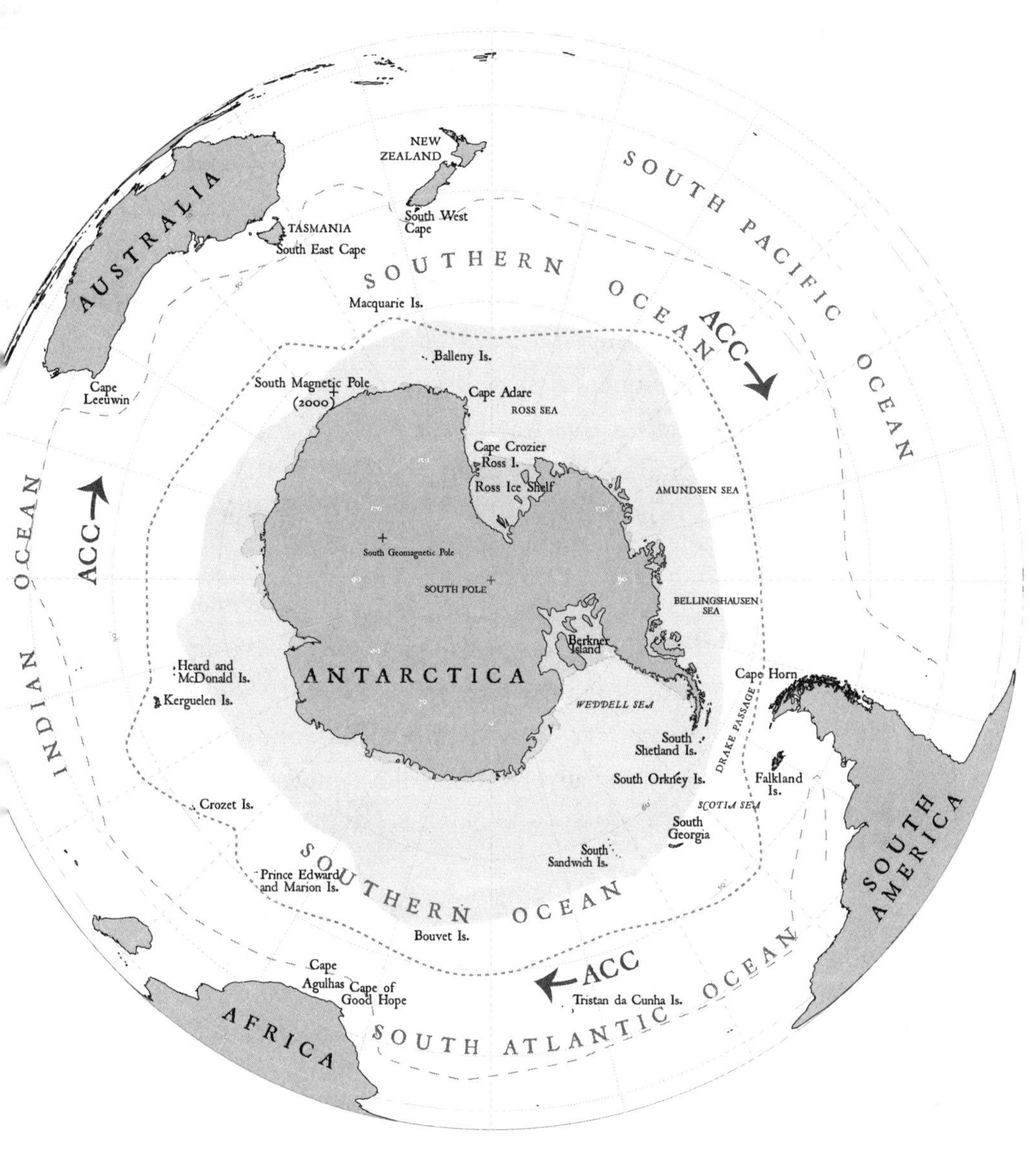

Antarctic Convergence

Subtropical Convergence

Extent of sea ice (maximum)

ACC Antarctic Circumpolar Current

the Southern Ocean. 'For by this time the mercury in the thermometer had fallen to 38[° Fahrenheit, or 3.3° Celsius]; whereas at the Cape it was generally at 67 [19° Celsius] and upwards.'[13]

Once we enter the convergence we may sight our first tabular icebergs setting out on their own voyages, caught in the thrall of the West Wind Drift having calved from Antarctic glaciers and ice shelves. 'Left to their own course', one writer optimistically observed in 1974, 'Antarctic bergs bother no one, drifting well outside normal commercial shipping routes. Once the big bergs calve off the ice shelves, they travel with the circular currents in the southern oceans and eventually melt.'[14]

Somewhere beyond latitude 55° South we are likely to encounter great expanses of compacted sea ice. Finally, we reach the ice shelves that skirt about 75 per cent of continental Antarctica. In the Weddell and Ross seas, which lie on either side of West Antarctica, the water becomes heavy as salt leaches out of the ice shelves, forming waterfalls below the ocean surface that plunge up to 2 kilometres into the abyss.[15] This cold, dense water known as Antarctic Bottom Water can take centuries to flow northward into the global oceans.

Finally, we turn back from Antarctica to the warmer latitudes, mingling with the neighbouring waters in the basins of the Atlantic, Pacific and Indian oceans. The complex layering and movements of warmer and colder waters is part of the slow dance of ice and water, heat and salt as the Southern Ocean converges with the other major oceans of the world, redistributing heat around Earth and influencing its climate. We have traversed the Southern Ocean from north to south, catching glimpses of the extraordinary power and complexity of its movements as it travels on its epic journey around the planet.[16]

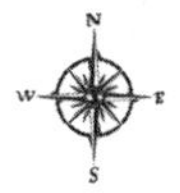

Ideas about how the Southern Ocean moved around the globe emerged surprisingly early in the history of southern maritime exploration. Mariners were well versed in the lores of surface currents along northern sailing routes by the seventeenth century, but Edmond Halley was the first to record the powerful eastward-flowing circumpolar current of the Southern Hemisphere, making observations from the deck of his British warship, the *Paramore*, in 1698.[17] An astronomer and mathematician with a particular interest in ocean winds and currents, Halley theorised that wind was really just another type of current, created by the warming action of the sun on air and water rather than by the motion of Earth itself, as claimed by earlier philosophers. To demonstrate, he published the world's first meteorological chart, showing the distribution of prevailing winds over the world's oceans. As the second astronomer royal at Greenwich at the time, he was also involved in attempts to establish a method of calculating longitude at sea.[18] A year later he ventured into the Southern Ocean in an attempt to discover the legendary southern lands, navigating into the ocean between the Strait of Magellan and the Cape of Good Hope and making magnetic and astronomical observations as well as recording weather, currents, flora, fauna and the colour of the sea. He described entering a region of thick fog and experiencing a dramatic drop in temperature, indications that he had crossed into the Antarctic Convergence, and he reached his furthest south at latitude 52° 24′ South where he sighted his first iceberg.[19]

During the early eighteenth century scientific interest was drawn to the movements of the winds rather than of the oceans.

In 1757 Sir Benjamin Thompson, Count von Rumford, an American-born British physicist with a fascination for the properties of heat, conducted some experiments with salt and water. Rumford observed that if enough salt was added to heated water, the water's density increased as it cooled. He speculated that sea water at freezing point would be dense enough to overturn all the way to the bottom of the sea floor.[20] The freezing surface water found off the coast of Brazil, he suggested, originated in the colder southern regions and flowed northward via deep-ocean currents. Contrary to the prevailing belief that heat was a liquid form of matter, Rumford was proposing that it was actually a form of motion.

> On being deprived of a great part of its Heat by cold winds, [water] descends to the bottom of the sea, [and since it] cannot be warmed *where it descends*, as its specific gravity is greater than that of water at the same depth in warmer latitudes, it will immediately begin to spread on the bottom of the sea, and to flow towards the equator; and this must necessarily produce a current at the surface in an opposite direction; and there are most indubitable proofs of the existence of both these currents.[21]

At the time of Cook's second voyage into the Southern Ocean, in the 1770s, nearly all the major surface currents of the world had been recorded, but the nature of the oceans' deeper circulation remained a mystery.[22] Cook's two astronomers, William Wales and William Bayly, were given the unenviable task of venturing out in a small wooden boat amidst the sea ice. Their task, as instructed by the Board of Longitude, was to measure the saltiness of the sea and the temperature of the water at different depths. Elsewhere, the ocean water was always a few degrees colder at depth. To their surprise, they found the deep water between 180 and 290 metres to be slightly warmer than the water at the surface.[23] Cook reached the conclusion that currents in the Southern Ocean flowed in a northerly direction and were responsible for bringing the ice from the south, but he did not see any connection between the changes in air and sea temperatures and the circulation of ocean currents.[24]

In 1819–21, Alexander I authorised the first Russian circumnavigation of the globe to the south polar region, led by naval officer and cartographer Fabian Gottlieb Thaddeus von Bellingshausen, aboard the *Vostok* and the *Mirnyi*. Bellingshausen's orders were to advance as far as possible towards the South Pole and investigate the polar currents and the sea temperatures and salinity in different regions and depths, to discover if there were differences in the gravity and density of the water. He was the first to use tow-nets night and day, capturing small drifting animals – and making the first recorded sighting of Antarctic krill – that came to the surface at night to feed and descended in daylight.[25] Along the way he surveyed South Georgia and established that Sandwich Island, first identified by James Cook, was in fact an archipelago of smaller islands.[26]

The British Antarctic explorer James Clark Ross made the first measurements of the tides of the Southern Ocean, during his surveying voyage to the Antarctic region with the *Erebus* and *Terror* in 1839–43. While his main aims were to measure Earth's magnetic field in the Southern Hemisphere and to establish the position of the South Magnetic Pole as part of the Magnetic Crusade, he was also instructed by the Royal Society to measure tides, temperatures, currents and ocean depths and to collect samples of water from different depths.

By the time Matthew Fontaine Maury published his book *The Physical Geography of the Sea*, in 1855, the nature of deep-ocean currents had become a subject of intense speculation amongst natural philosophers. Maury was not just a cartographer of the sea; he was an evangelist for the ideas of natural theology. Indeed, his aim was not simply to present the current state of knowledge about the oceans for navigational purposes but to inspire others with his passion for the 'wonders of the great deep'. He wrote,

> In the beautiful system of cosmical arrangements and terrestrial adaptations by which we are surrounded, they [ocean currents] perform active and important parts; they not only dispense heat, and moisture, and temper climates, but they prevent stagnation in the sea; and by their active circulation, transport food and sustenance for its inhabitants from one region to another, and people all parts of it with life and animation.[27]

Far from 'running hither and thither', he believed, the currents' movements were guided by some law of nature. Drawing on reports from voyages to the Antarctic region, Maury speculated that warm water flowed southward from the equator through the Indian Ocean, while an ice-bearing current flowed as far north as latitude 40° South, 'wending its way from the Antarctic regions with supplies of cold water to modify climates'. The circulation of these immense volumes of water did not in itself explain climatic conditions, but it seemed to Maury that the key to understanding weather lay in the embrace of the 'two oceans of air and water'.

> Our planet is invested with two great oceans; one visible, the other invisible; one underfoot, the other overhead; one entirely envelops it, the other covers about two thirds of its surface … It is at the bottom of this lighter ocean where the forces which we are about to study are brought into play … at the meeting of these two oceans … They are both in a state of what is called unstable equilibrium; hence the currents of one and the winds of the other.[28]

It was a revolutionary idea: that the ocean and atmosphere were entangled both vertically and horizontally, 'reaching from the sun and moon', as historian Michael Reidy put it, 'through the trade winds and magnetic properties of the atmosphere, all the way down to the tidal currents on the ocean's surface'.[29]

When scientists on the *Challenger* expedition, in the 1870s, observed that sea floor waters off Brazil were almost freezing despite their proximity to the equator, they speculated that the water must have travelled northward to the equator from the Antarctic region via a deep-ocean current. They also observed that offshore summer winds sucked upward ice-cold water from the abyss of the Indian Ocean, chilling the surface temperature along East Africa's coastline and taking dry conditions to Australia's west.[30] It proved that when very cold deep-ocean water flowed along the western coastline of a warm landmass, it produced dry conditions. John Walter Gregory described the phenomenon at a meeting of the Australasian Association for the Advancement of Science in 1904. The *Challenger* expedition, he said, had produced conclusive evidence of the 'oceanic control of the climate of the lands in the Southern Hemisphere'.

The ability to predict, even control, climatic conditions in the Southern Hemisphere was foremost in Gregory's mind at the meeting. Australia was in the midst of a catastrophic eight-year drought at the time, and it cast a deep shadow over the recently federated nation. Australia's entire wheat harvest had failed, and sheep and cattle lay dead in their millions. Gregory urged his audience to support the establishment of a 'United Meteorological Service for Australasia' modelled on the successful Indian service.

> If, as was once thought, ocean currents flowed on fixed and permanent courses, like rivers on the land, then their effects upon the climate of the adjacent lands should also be unchanging; but as the air circulation varies with the season, it is only natural that the oceanic circulation should also vary at different times of the year.

If the 'weather be determined by the remote upwelling of water from the deep seas', he declared, then the fledgling science of meteorology was ill equipped for the task without an understanding of the unpredictable currents and winds of the Southern Ocean that regulated it. It would be the 'beneficent sea [that] restores the equilibrium which is necessary to our existence'.[31] Some observers were highly amused by Gregory's approach to weather forecasting. One journalist described to his Sydney readers how the meteorologist of the future would be 'a blue-spectacled person, who [would] take his observations out in an open boat in the Southern Ocean, measuring currents like a skilful housewife with a Christmas pudding on hand, and taking temperatures like a busy doctor in the typhoid season'.[32]

By 1908 Australia had its own Bureau of Meteorology, providing the nation with a weather forecasting service and institutional support for scientific investigations into the relationship between ocean and atmosphere in the Southern Hemisphere. 'There is probably no country in the world', proclaimed one Melbourne newspaper, 'not excepting even the United States of America[,] which is so vitally affected by its varying weather conditions as Australia. The general character of a season may make all the difference between prosperity and financial strain to its people … Australians have every incentive to the study of meteorology.'[33]

But the task of solving the enduring mysteries of ocean circulation and its influence on climate was just beginning. Data collected from the Southern Ocean by the *Challenger* scientists had yielded only a very basic understanding of its water masses and their movements. They had detected cold water of low salinity at the surface of the ocean near Antarctica and presumed that similar water found near the equator had originated from those higher latitudes, but the mechanism by which such large bodies of water could circulate over such long distances remained an enigma. A series of Antarctic expeditions around the turn of the century provided scientists with greater opportunities to study the movement of the water masses at first hand.[34] Armed with more-accurate deep-sea thermometers than those available to the *Challenger* scientists, they confirmed that the Southern Ocean had a distinctive layering of cold surface water over warmer deep water over cold bottom water.

By the end of World War I, scientific interest shifted from the Southern Ocean's flora and fauna, which had preoccupied the *Challenger* expedition, to the nature and circulation of its waters. The development of new technology for detecting submarines had given an unexpected stimulus to oceanographic research and, over the next few decades, increasingly sophisticated underwater technologies would enable scientists to learn more about the movements of water currents and how they were shaped by the topography of the underlying sea floor. In 1923 the German meteorologist Wilhelm Meinardus examined the results of the *Gauss* expedition conducted in 1901–3 and proposed that the transition between the cold polar waters from the south and the warmer waters from the north, as recorded by Halley, Cook and other explorers, occurred where cold polar water sank beneath warmer surface water. He concluded that the cold water kept moving northward in the form of a deep-level current. The Meinardus Line, as it became known, described for the first time the Southern Ocean's most distinctive oceanographic feature: the Antarctic Convergence.[35]

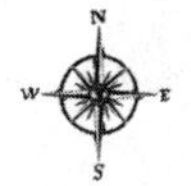

Upwelling adds another dimension to the ocean's remarkable qualities. Along the Southern Ocean's remote southern coastlines and islands, some of the strongest winds on Earth push surface waters out to sea and draw up cold, nutrient-rich waters from the deep. These areas of upwelling are often likened to a window between Earth's atmosphere and its abyss, enabling the ocean to absorb carbon from the atmosphere.[36] They also host microscopic phytoplankton and zooplankton that thrive in the cold polar waters, providing the pasture for swarms of krill, which in turn attract a throng of albatrosses, whales, penguins, fish, seabirds and seals – crabeater, leopard and Ross.[37]

Upwellings make the Southern Ocean bloom. They also underpin the world's fisheries; 25 per cent of all ocean fish caught are found in five upwellings that collectively cover just 5 per cent of the ocean surface. Wherever upwellings occur, as Rachel Carson put it, they 'set off orgies of devouring and being devoured'.[38] Each year some of the greatest concentrations of baleen whales in the world gather to feed in the ocean's waters, including the largest animal ever to exist on Earth, the blue whale (*Balaenoptera musculus*). An adult blue whale can measure up to 30 metres long and weigh over 160 tonnes.[39] Its heart alone weighs as much as a motor vehicle. However, in a quirk of nature, it feeds almost exclusively on krill which grow to a length of just 6 centimetres. An adult blue whale can consume up to 40 million krill each day.[40]

Baleen whales are distinguished from toothed whales by the baleen plates in gums along each side of the whale's upper jaw, which are designed to strain food from the water. Baleens including the blue, pygmy blue, fin, sei, southern right, minke and humpback, as well as toothed sperm whales, are renowned for their seasonal migrations between their feeding grounds around Antarctica and their breeding grounds off Australia, New Zealand, South America and South Africa. It was these migrations that lured whalers to the Southern Hemisphere oceans. By the late nineteenth century whale products were in demand in the factories and households of Europe and North America. Whale oil had become a popular fuel for lamps: the viscous sperm whale oil was considered one of the finest for this use, as it burned slowly without a strong odour. It was favoured also for manufacturing candles, soaps, cosmetics, paints and some medicines. Whalebone, or baleen, was put to domestic uses as struts for corsets and umbrellas. As the numbers of whales in Arctic waters declined as a result of unregulated commercial whaling, privately sponsored whalers had wasted no time in heading for the promising new southern oceans. The story of whaling in the Southern Ocean followed the same pattern as that of sealing. As one species was hunted to the brink of extinction the whalers turned to another, converging on areas where their prey gathered to breed or feed, then slaughtering and extracting precious oils until the beaches were silent and the waters empty.[41] Humans had become the top predator in the Southern Ocean by a nautical mile.

One species of large baleen whale hunted by whalers along the southern coastlines of Australia was the southern right whale (*Eubalaena australis*). Southern right whales spend the year circumnavigating the oceans of the Southern Hemisphere between latitudes 16° South and 65° South. In winter months – between May and October – several thousand follow ancestral migration routes northward from the Antarctic waters to nurse their young in favoured bays such as Doubtful Island Bay and Israelite Bay in southwestern Western Australia, and further east at the Head of Bight in South Australia. Some take shelter in the southern coastal bays. Others follow the eastern Australian coastline as far as northern New South Wales. The species is distinguishable from other baleen whales by its broad back with no dorsal fin, a long arching mouth beginning above the eye, and an enormous head with small rough patches of skin.

It is likely that these docile, slow-swimming creatures were given their English name because they were the best, or right, whale to kill. Certainly, their habit of lingering in sheltered coastal waters made them easy targets for crews in small boats. Baleen whales are easily distinguished from toothed whales by their baleen plates and paired blowholes, through which they exhale a column of vapour before inhaling to submerge. Whalers would call out 'Thar she blows!' to alert the crew when they spotted a baleen whale before it dived. Once harpooned, the whales' massive fat-rich bodies would rise to the surface and float, making it a relatively simple task to tow the carcasses to shore, where they would be cut up for meat and boiled to release large quantities of their highly prized oil.

The Southern Ocean whaling industry not only delivered great wealth to commercial enterprises in the Northern Hemisphere; it also had a long-lasting impact on the remote southern coastlines and islands of the Southern Hemisphere. The first whaling station in the Australian colonies was established in 1806, at Ralphs Bay in Tasmania. Early whaling was conducted from vessels either at sea or close to southern coastlines and islands where whales could be harpooned from small whale catchers and towed to shore-based processing factories. Some whalers combined pelagic (open-sea) whaling and bay whaling.[42] By 1841 there were an estimated 58 shore-based whaling stations in the Derwent estuary and along Tasmania's southern and eastern coasts.[43] The whales around these coasts were abundant, and harvesting their oil and baleen plates made a few men very wealthy. Hobart was ideally located in the path of the Roaring Forties, offering a convenient resupply point and accommodation as well as a port for bay whaling and safe harbour for those voyaging out into the wild winds and seas of the Southern Ocean. It rapidly transformed into a major hub for sealing and whaling vessels. No fewer than 300 vessels registered at the port between 1816 and 1823, and almost half were from the northern whaling hub of New England on the Atlantic coast of the United States, from where the American sealing and whaling trade mainly operated. The sealing vessels were based in Connecticut and the whalers in New Bedford, Massachusetts.[44]

In 1840 ship's surgeon WH Leigh wrote,

> The shores of Van Diemen's Land [Tasmania] are the resort, at certain seasons, of the whale, which proceeds thither for the purpose of calving in the numerous bays with which the island is indented. The whale is of the black species; and at the various bays there are whaling establishments belonging to the merchants of Hobart Town or Launceston.[45]

The settlement of Portland on the southern coast of Victoria had its origins in the 1820s, when itinerant sealers arrived in search of southern fur seals. Some stayed on, combining small-scale whaling and grazing operations. William Dutton, a seafarer originally from Hobart, first landed at Portland Bay in 1828, on the first of several sealing trips for John Griffiths of Launceston. After a period of sealing, he decided there was more money to be had in whaling and returned to Portland Bay to establish a whaling station there. By 1836 he was dividing his time whaling between Portland Bay and Recherche Bay, in Tasmania. Later, he took up 260 hectares of grazing land near Portland.[46]

When the Scottish surveyor and explorer Thomas Mitchell visited Portland in 1836 on one of his extended surveying forays in southeastern Australia, he found large gatherings of whales and a prosperous whaling and pastoral business owned by the brothers Edward and Frank Henty. In the previous year they had shipped 700 tonnes of whale oil; according to Mitchell, whales were so plentiful that when the casks were full any excess oil would be emptied into pits.

Whaling also became a form of local entertainment for the colonists of Portland. Rival teams competed to harpoon the greatest number of whales in the bay, while spectators watched the crews battle with the 'sea giants' from vantage points on the clifftops and specially built grandstands. Whale catcher boats, each with a bow painted a different colour, would be drawn up in line on the beach ready for instant launching. During one race, as Edward Henty's nephew Richmond recalled, a whale was being pursued by no fewer than 33 boats while the oarsmen were urged on by their steerers. 'Harpoon after harpoon is rapidly but somewhat wildly thrown from the nearest boats, but the fish escapes as, turning again, it dives deep below and under all the boats, racing back the way it came.'[47]

Indigenous people were employed in the sealing and whaling industries of southern Australia and New Zealand, sometimes moving between the various hunting grounds. Men from Australian coastal communities were often sought after to work on whaling boats because of their skill in throwing harpoons, and in New Zealand the Māori were regarded as excellent sailors and whalers. Early colonial records suggest that some Aboriginal Australian communities considered whale meat a delicacy, and while they did not actively hunt whales, they would harvest the meat of a stranded whale. In 1881 James Dawson observed that the Peek Whuurong people in southwestern Victoria would bury the flesh of a stranded whale for later consumption. In her study of Aboriginal Australian whalers and sealers in the oceans of the Southern Hemisphere, Lynette Russell cited several instances in which colonial observers recorded large gatherings of Aboriginal people to feast on stranded whales, indicating that such events had a long history in southern Australia and provided an important food source as well as a focus for ceremonial and social occasions.[48] A painting made by the convict artist Joseph Lycett in around 1817, for example, depicts small family groups gathered on a beach near Newcastle, on the New South Wales coast, to cook and eat the flesh of a beached whale.[49]

Single strandings tended to involve baleen whales, such as the humpback, southern right and minke. Mass strandings, on the other hand, were more likely to be of toothed whales, such as the sperm, pilot and pygmy sperm; these were particularly common along Tasmania's wild western shores. The whales were not only a source of food, as Russell noted. Strandings were also associated with ceremonial gatherings and played a significant spiritual role in the lives of the Aboriginal people of southern Australia. The anthropologist Norman Tindale observed in the 1930s that whales served as totems for some groups. Those who belonged to the whale totem would not eat whale meat, and if a whale seemed to be struggling near shore, they would sing to it to encourage it to avoid the shallows and head back into the ocean.[50]

Shore-based whaling stations attracted groups of local Aboriginal people, who would gather to collect pieces of unwanted whale flesh. In the 1830s, for example, Leigh recorded his impressions of a whale kill during a visit to Encounter Bay, in Narrinyeri country.

> The natives, who had assembled in numbers upon the beach, immediately pounced on the offal, cutting off lumps as large as they could move under, and pushing their spoil before them as they swam ashore … These pieces were removed to a convenient place for the entertainment, and a party of 'running footmen' were dispatched to deliver invitations to the neighbouring friends to come and eat whale with them![51]

Whalers also ventured further south, into the polar region, following the large baleen whales on their summer migration to feed in the waters around the Antarctic coastline as the sea ice receded. When James Clark Ross published an account of his four-year voyage in the Southern Ocean in 1847, he included observations about not only the nature of the ocean's depths but also the abundance of its whales. At latitude 71° 50′ South he counted 30 large whales around the vessel at one point and many more during the course of the day. 'Hitherto,' he noted, 'beyond the reach of their persecutors, they have here enjoyed a life of tranquillity and security; but will now, no doubt, be made to contribute to the wealth of our country, in exact proportion to the energy and perseverance of our merchants.'[52]

Charles Enderby was one of those merchants who sought to live up to Ross's claims. Ross met with him to encourage his financial backing for a colony on the Auckland Islands, south of New Zealand, that would service the influx of whaling ships and process their catches. With images of a verdant agricultural landscape and prosperous new settlement in mind, in 1849 Enderby formed the Southern Whale Fishery Company and promoted the islands to young married couples as a *terra nullius* that offered 'a very rich virgin soil' and a bracingly healthy climate.[53] In truth the Aucklands, in common with the other subantarctic islands of the Southern Ocean, are relentlessly windy and wet, with rugged mountainous terrain and boggy peat soils. Sir George Gray, the governor-in-chief of New Zealand, described the weather as 'continuously vile' when he visited the islands in 1850.[54]

By the time the whaling company was formed, the islands were familiar territory to the Enderby family. Charles's grandfather Samuel Enderby had founded Samuel Enderby & Sons in the mid-eighteenth century to exploit sealing and whaling fisheries in both hemispheres. When Enderby vessels began voyaging into the Southern Ocean around Cape Horn, their masters were encouraged to combine geographical exploration of the ocean with the discovery of new whaling grounds. Ships sailing between Cape Horn and Australia around the southern coastline of New Zealand were likely to pass the Aucklands along the way and one of Enderby's captains, Abraham Bristow, was the first European to sight them, in 1806, aboard the British whaling ship *Ocean*. He returned the following year to claim the islands for Britain, naming them Lord Auckland's after his father's friend William Eden, 1st Baron Auckland.[55]

The British were not alone, however. When the first of the ships carrying Charles Enderby's new settlers arrived, in 1849, they were greeted by a group of Māori and Moriori people, who had relocated there from the Chatham Islands eight years earlier. Enderby tolerated their presence providing they made no claim to the land, and they settled on the northeastern point of the island and managed to cultivate small plots and raise pigs that had been left behind by one of Enderby's ships.[56] Enderby's whaling settlement went ahead, but it survived for just three years before yielding to the island's inhospitable nature. Enderby abandoned it to the elements, and to the Māori and Moriori, having caught just one whale.[57]

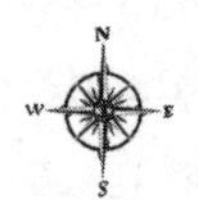

South Georgia is long and narrow, bending around the Scotia Arc at the edge of a tectonic plate, its deeply indented coastline pummelled into shape by the circumpolar ocean. It is an exposed part of the submarine ridge that connects Tierra del Fuego and the Andes of South America with the Antarctic Peninsula. It is just 160 kilometres long and 32 kilometres wide and is crowned by a ridge of forbidding mountain peaks that are perpetually covered with snow and ice. The long black ridge seems to hold the island together.

Several early sailing captains reported sighting land in the vicinity of South Georgia, and Cook charted it in 1775, claiming it for Britain and naming it the Isle of Georgia, in honour of George III. Over the next century the island was a regular haunt for British and American sealers, who set up temporary camps for the seasonal slaughter of fur seals. As unregulated sealing progressively exterminated the fur seal colonies, commercial interests turned to processing oil from the other abundant mammals of the region, targeting elephant seals which bred in huge numbers on many of the subantarctic islands, and the large migratory southern right, blue and fin whales that converged on the rich veins of nutrient-rich water upwelled from the deep.

In 1894 Carl Anton Larsen, a Norwegian explorer, observed the abundance of whales around South Georgia during an Antarctic voyage to investigate the potential for hunting southern right whales. He visited again in the winter of 1902, in command of the *Antarctic* which was carrying Nils Otto Gustaf Nordenskjöld's Swedish Antarctic Expedition, and remarked on the enormous numbers of large whales in the waters around the island.[58] Aboard the *Antarctic*, J Gunnar Andersson described his first impression of South Georgia: 'Southwards, a magnificent Alpine country, illuminated by the rising sun, rose slowly from the sea; there were mighty fells with snowy crowns and with sharp, uncovered teeth, around the valleys through which enormous, broad rivers of ice came flowing to the sea.'[59]

The *Antarctic*'s voyage was financed by Svend Foyn, a Norwegian whaling captain, who had invented the steam-powered whale catcher and patented the grenade harpoon gun. The exploding harpoon meant that whale hunting became faster and more efficient and, together with his other invention, enabled whalers to pursue the larger and faster species of whales. The expedition visited Tristan da Cunha, Prince Edward Islands, the Crozet Islands, the Kerguelen Islands, the Balleny Islands, Campbell Island and Possession Island before landing at Cape Adare, making the first confirmed landing on the Antarctic continent. The expedition ended dramatically, however, when the vessel failed to reach its base on Snow Hill Island, off the east coast of the Antarctic Peninsula, and was soon trapped in sea ice and crushed. Larsen, with the expeditioners and ship's crew, was forced to over-winter on Paulet Island. They were eventually rescued by an Argentinian naval ship which sailed to Buenos Aires, and there Larsen took the opportunity to progress his plans for a shore-based whaling station at Grytviken, on South Georgia. With Argentinian financial backing and armed with two transport ships, a prefabricated factory and around 60 men, Larsen's whaling venture boomed. His success attracted other whaling enterprises to South Georgia and the nearby South Orkney, South Shetland and South Sandwich islands, and those remote groups were soon at the centre of the most profitable whaling grounds in the world.[60]

When the American ornithologist Robert Cushman Murphy visited South Georgia, the southernmost outpost of the British Empire, aboard the whaling vessel *Daisy*, in 1912, he found Larsen to be the undisputed 'king of modern whaling', clearly enjoying the fruits of his whaling empire. During his stay Murphy chronicled both the gory business of shore-based whaling and the trappings of comfort and civilisation in its midst.[61] The settlement accommodated 500 families and boasted a 'fully-fledged Empire Post Office' as well as a Lutheran chapel, while Larsen's own residence was adorned with palms and a conservatory of plants complete with singing canaries and a portrait of the Norwegian king. By comparison, the conditions that the ordinary whalers endured on South Georgia were so grim that the island was referred to as 'the slum of the Southern Ocean'.[62]

Out on a whale catcher, Murphy noted the profusion of whales on the northeastern coast of the island, where the cold polar waters converged. There he witnessed the 'big-scale butchery of modern whaling' and recorded how, at one point, harpoon guns were banging continuously from no fewer than 11 of Larsen's whale catchers as they converged on the whales' feeding grounds. Of all the Southern Ocean's whales, the giant blue whale was the species that whalers prized the most. When the *Fortuna*, one of Larsen's vessels, began its pursuit of a blue whale, Murphy was entranced by the animal's 'marvellous grace'. It was not 'shapeless with all the firmness and streamline of the body gone', as he had once thought of the bloated carcasses he saw washed ashore on Long Island beaches. Rather, he saw its 'sheer beauty, symmetry, utter perfection of form and movement' and appreciated the immense size of the animal, which dwarfed the hull of the *Fortuna*. He imagined 'this magnificent blue whale' in a noble battle, as 'shapely as a mackerel, spending its last ounce of strength and life in a hopeless contest against cool, unmoved, insensate man'. Indeed, the whale's 'utter perfection' could not save it from the determination of the *Fortuna* crew, who were 'armed with devices worse than those of the serpent'.[63] Whales were not necessarily easy prey, as an official Australian observer aboard the Japanese factory ship *Hashidate Maru* attested in 1947. In words reminiscent of a military campaign, he reported that a blue whale being pursued by the catcher had managed to evade the harpoon. It had clearly been chased before, he noted, and knew the 'method of evasion too well'.[64]

Murphy's whaling excursion ended when a fog descended over the battle scene like a theatre curtain at the end of a play. The slaughter of such a majestic creature had seemed incongruous to him, but he could not help but admire the ruthless efficiency of it all. Within just two days the 60 whales he had counted on the slipway had been turned into oil and bone meal for fertiliser. Meanwhile, an evening walk along the beach nearby revealed an abundance of whale carcasses and an extraordinary assortment of other dead animals. 'It is an infinitely sheltered and soft and peaceful evening here', he wrote to his wife, 'despite the acrid whaley smell.'[65]

By the summer of 1912–13 South Georgia had six shore-based stations operating at full capacity. The lack of regulation and the alarming decline of whale 'stocks' were matters of concern to local authorities. In addition to the shore-based processing facilities, there were factory ships anchored in bays along the Antarctic Peninsula and in the South Shetland, South Sandwich and South Orkney islands, kept busy with whales harpooned from 62 catcher boats. In that single season, 10 760 whales were taken.[66] William Allardyce, who served as the governor of the Falkland Islands between 1904 and 1914, recognised the need for some form of restraint on whaling in the surrounding waters and advised the Colonial Office in London to introduce a licensing system to limit the number of shore-based whaling stations, floating factory ships and whale catchers. The threatened southern right whale was given full protection under the resulting regulations.[67] The killing of female whales with calves was banned under the new system, which also set quotas for catches and required whalers to process all parts of each whale. Until then whalers had been taking little more than the blubber and baleen, leaving most of the carcasses to rot on the beach.

In 1914 the British Colonial Office dispatched Major Gerald EH Barrett-Hamilton to South Georgia to gather at first hand information about whether the whales of the Southern Ocean were 'in danger of extermination'. He was accompanied by a taxidermist tasked with collecting whale specimens for the British Natural History Museum. Barrett-Hamilton died while in South Georgia, but his work was continued by Percy Stammwitz, who produced annual reports on whale numbers and furnished the Natural History Museum with specimens for its collection.[68]

The declaration of war in Europe intervened in this conservation work but it seems that, rather than mitigate the pace of whaling in the Southern Ocean, the conflict served to seal their fate. The fat of baleen whales is a rich source of glycerol, or glycerine which, when mixed with nitric and sulphuric acid, produces nitro-glycerine, used extensively for the manufacture of explosives during the war. Whale oil was also issued to soldiers as a preventative treatment for trench foot.[69] In 1915–16 alone, 11 792 whales were killed around South Georgia. As the author Philip Hoare observed, the scale of human slaughter on the Western Front seemed to sanction the wholesale slaughter of whales in the world's oceans. But the greatest blow to Southern Ocean whales was still to come. In 1925 steam-powered factory ships largely replaced shore-based whaling stations. First developed in 1905 by Christen Christensen for use around the South Shetland Islands, factory ships were ruthlessly efficient. Each could spend extended periods away from shore. Once harpooned, whales were hauled onto deck via a stern ramp (first fitted in 1925 on the Norwegian factory ship *Lancing*) for immediate processing.[70] The result of this new style of pelagic whaling was a far more flexible and rapacious industry and an even greater devastation wrought on the whales of the Southern Ocean. In 1926 the London *Times* declared that whales would soon cease to exist unless action was taken to protect them.[71] Within five years the Southern Ocean whales were being pursued by 41 factory ships and 205 catchers, and the numbers of whales killed each year soared, from 8448 in the 1920–1 season to 14 219 in 1925 and 46 039 in 1937–8. In the 1930–1 season alone, the total whale catch included a record 29 400 blue whales.[72]

Alarmed by the rapidly shrinking whale populations in the seas around South Georgia and the subsequent threat to its whaling industry, the British Government launched one of the most ambitious and significant studies of whale biology and behaviour to have been attempted at that time. The Discovery Investigations comprised a series of scientific voyages, undertaken between 1925 and 1951.[73] The expedition scientists conducted natural history research in the Southern Ocean alongside the British whaling fleet. They were present not for the purpose of killing whales, as the *Discovery*'s chief zoologist Alister Hardy stressed, but 'to find the facts that might help in their conservation'.[74] The voyages were paid for by a levy imposed on the whaling and sealing industries by the British Government and were organised by the Interdepartmental Committee for the Dependencies of the Falkland Islands (often called the Discovery Committee). A report by the committee to the British Parliament in 1920 had recommended that whales be marked and their food sources investigated, that commercial activities be controlled and that whaling stations have a resident zoologist to monitor their proceedings. The expedition's primary objective was to gain a better understanding of whale biology as well as the natural environment on which they depended, so as to ensure the future of the whaling industry in the region. The quest for scientific knowledge merged seamlessly with the quest to protect Britain's economic interests and save the whale 'stocks' from annihilation.

The committee invited the South Australian geologist and Antarctic explorer Douglas Mawson, who had first come to public attention as the leader of the Australasian Antarctic Expedition in 1911–14, to organise and lead the first voyage. It supplied him with Robert Falcon Scott's old wooden-hulled sailing vessel the *Discovery*. Equipped with auxiliary steam engines, *Discovery* sailed south from London on 24 September 1925. It was joined a year later by a smaller purpose-built whale catcher, the *William Scoresby*, named after the nineteenth-century British Arctic explorer, scientist and clergyman. At South Georgia the scientific team established a marine biological laboratory in Discovery House, on King Edward Point, close to the Grytviken whaling station where the carcasses of whales could be examined.[75] They also began a program of marking 5219 whales with identifying darts in order to collect data on breeding and migration patterns and undertook more general studies of krill, the whales' main source of food.[76]

It was soon clear, however, that the old square-rigged sailing vessel *Discovery* was not up to the task of long-distance voyaging in the Southern Ocean. The committee replaced it, in 1929, with one of the world's first purpose-built oceanographic research vessels, the fully steam-powered Royal Research Ship *Discovery II*.

The new vessel was fitted out with two scientific laboratories and gear for retrieving samples from the depths of the sea. Over the next decade both ships zigzagged around the circumpolar ocean, taking measurements and collecting samples of the whales' habitat and food supply for scientists and museums back home. Mawson saw five enormous Norwegian whaling factory ships in a single week in the summer of 1930–1. He recalled 'a constant stream of these monsters' – 150-tonne whales – arriving on the flensing decks to be reduced to oil and other products, each one processed in less than an hour.[77]

Carl Anton Larsen had left the Grytviken station at the start of World War I but, after limited success with farming ventures in Norway, had returned to Antarctica with the factory ship *Sir James Clark Ross* to hunt whales in the Ross Sea. Larsen had died in 1924, while whaling.[78] In 1930 Mawson arranged for his vessel, on its maiden voyage, to take on coal from Larsen's old ship, and Gilbert Eric Douglas made use of the opportunity to inspect its operations at close quarters, as well as to participate in a whale chase. His logbook reveals the devastating efficiency of the new harpoon technology:

> At about 8.30 we manoeuvred into a good position and up came a Blue Whale, he spouted several times, the gunner trained his gun, the whale rose again prior to sounding, up went his huge back. Our gunner is intent on his aim, Bang! the harpoon flew out, Smack! the harpoon went right into the whale, up came his huge tail in a wild sweep. Boom! our explosive nose cap had gone off inside the whale. Away went the whale at a terrific speed, one hand now stood at the winch and braked on the heavy rope as it tore outwards,

> smoke rose from the rope, where it passed through another wooden break in the deck ... Our whale was nearly done, one last flounder and he lay still. Pigmy like man had mastered the largest living animal and this 100 tons of blood and energy was stilled for ever.[79]

Whaling had become a brutal industry by any measure, although whalers were not always deaf to the suffering of their prey. A Scottish whaler who worked on South Georgia recalled how whales would cry when the harpoon found its mark.[80] When naturalist and nature writer Francis Downes Ommanney visited the whaling operations at Grytviken during the 1920s as a member of the Discovery Investigations, some of the harpooners admitted to him that they did not really like their job, especially when their harpoon missed its target during a hunt. Sometimes, instead of exploding in the whale's backbone causing instant death, the tip exploded somewhere inside the animal. Ommanney later wrote,

> There is difficulty of imagining that this grotesque creature, plunging and wallowing at the end of the line, is a beast as sentient as a horse and, in its own way, as noble. Its habit of life in its unfamiliar element makes it impersonal and mysterious. What an outcry there would have been long ago, as Sir Alister Hardy has remarked, if herds of great land mammals, say elephant or buffalo, were chased in armored vehicles firing explosive grenades from cannon, and then hauled close at the end of a line and bombarded again until dead.[81]

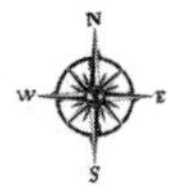

During the 1930–1 summer season record numbers of whales were killed in the Southern Ocean. Whalers from Britain and Norway caught 95 per cent of all whales harvested, with Germany, the Soviet Union, the Netherlands and Japan responsible for the remaining 5 per cent. Voices had been calling for restraint from the earliest days of whaling in the region, warning of depleted stocks and advocating greater regulation to ensure a future for the industry. A general survey of whaling undertaken in 1931, based on statistics provided by the whaling industry, showed the rapid decline in whale numbers since the introduction of pelagic whaling factories.[82] Amidst growing concerns amongst whaling nations that such indiscriminate harvesting would result in the extermination of whole species and hasten the demise of the industry, British and Norwegian whalers petitioned the League of Nations to impose restrictions on the operation of factory fleets. This may have been motivated by self-interest, but it signalled the beginning of co-operative international efforts to conserve whales in the Southern Ocean. The League responded by establishing an international bureau to monitor whale catches and made concerted attempts over the next decade to impose international restrictions on whaling operations. In 1931 the League negotiated the *Convention for the Regulation of Whaling* with 22 of the major whaling nations, the first international regulatory agreement on whaling, although whaling nations continued to debate the ways in which the industry should be regulated. It came into force in 1935 and Britain and Norway reached agreement on a closed season between December and March and a limit on the number of catchers that could be employed by a factory ship. As the world edged towards another war, several nations participated in an international conference in London and agreed to ban pelagic whaling for nine months of each year and prohibit killing of certain species and in certain areas of the Southern Ocean.[83]

World War II may have given the whales some breathing space, but industrialised whaling resumed apace at the end of hostilities, and another chapter in the long history of 'serial exploitation' of the Southern Ocean unfolded. One of the first acts of the United Nations after the war was to establish the International Whaling Commission, in 1946, with 14 member states. Two years later the *International Convention for the Regulation of Whaling* came into force, 'to provide for the proper conservation of whale stocks and thus make possible the orderly development of the whaling industry'.[84] By the time South Georgia's six shore-based stations ceased operations, the last doing so in December 1965, they had notched up a total of 175 250 whales. It had become increasingly clear that there were too few large whales remaining in the Southern Ocean for the southern whaling industry to be commercially viable.[85]

When Graham Chittleborough began his job for the Australian Government's Commonwealth Scientific and Industrial Research Organisation, in 1950, the CSIRO was involved in rebuilding Australia's whaling industry after the war, in order to meet manufacturers' demand for whale oil products such as margarine. By then private whaling operations were in full swing in Western Australia – at Cheynes Beach in Frenchman Bay, near Albany; and at Point Cloates, near Exmouth – as well as along Australia's east coast, at Tangalooma, in Queensland; Byron Bay, in New South Wales; and Norfolk Island. The government had established the Australian Whaling Commission to run its whaling operations at Babbage Island off Carnarvon in Western Australia. Its chief concern was to ensure that whaling remained a viable industry off Australia's western and eastern coastlines along which large numbers of whales, particularly humpbacks (*Megaptera novaeangliae*), migrate between their Antarctic feeding grounds and the warmer equatorial waters to mate and give birth each year. Their northern migration takes them offshore, but on their southward journey they track close to shore to allow their calves to rest in calm, shallower waters. According to Australian marine biologist Micheline Jenner, humpback whales make the longest seasonal migrations in the animal kingdom.[86]

Within two seasons it was evident to Chittleborough that the quotas imposed by the government were dismally inadequate to stop the decline of humpback whale numbers. By 1955 the Western Australian whaling stations accounted for about 60 per cent of the total number caught in Australia's coastal waters, but the whales being caught were getting smaller, and boats were having to travel longer distances to find the larger animals. In 1960 Chittleborough reported that the humpback population had declined to a point at which immediate action was necessary in order to avoid its complete destruction. After a mere decade the situation was so dire for humpback whales migrating from Antarctic waters that the International Whaling Commission determined that Australia's shore-based humpback whaling operations should cease altogether. Nevertheless, commercial whaling vessels continued to ply the open seas of the Southern Ocean, although a combination of dwindling whale populations and emerging public opposition to commercial whaling meant that their days too were numbered.

On the Bunda Cliffs, at South Australia's Head of Bight, the wind seems to come from all directions at once. The crumbling limestone cliffs define the edge of the Nullarbor Plain, stretching for about 100 kilometres between Western Australia and South Australia. It is the Great Australian Bight that gives the southern coast of Australia its distinctive arc. The winds buffeting the cliffs have gradually exposed the shells of an ancient seabed, a reminder that it formed when Australia and Antarctica split apart and opened the way for the Southern Ocean to flood in. Land eventually emerged from the ocean, and the longest continuous limestone cliffs in the world were born. Each year the cliffs provide the dramatic backdrop to a spectacular gathering of southern right whales who migrate there with their calves to feed on the upwelled nutrients and nurture their young.[87]

The ocean is part of the ancestral country of the Mirning people of the Great Australian Bight and Nullarbor (Ngargangurie), who maintain their traditional language, Ngandatha. The whale is the totem of the Mirning people, and they share a common ancestor with the great white whale Jeedara who came from the Milky Way with the stars known as the Seven Sisters. Jeedara hid in the rocks to escape the Seven Sisters, but they discovered his hiding place and pursued him. In the chase that followed, Jeedara flung himself around and hit his tail against the rocks, forming the Bunda Cliffs and the web of tunnels and caves that lie deep beneath the Nullarbor. Each year, the whales that migrate to the Bight are honouring Jeedara's great journey from the Milky Way. Bunna Lawrie, the Mirning's whale song man, story teller and keeper of the whale Dreaming songlines, was born at the Koonibba Mission at the edge of the Nullarbor. 'The Great Australian Bight', he said, 'is the greatest whale nursery on this planet. The whale story where I come from is my university, my school. It's the place where our beautiful southern right whales come to calve their young, to teach their young to travel on the next journey.'[88] For Bunna's tribe, the memory of the slaughter of the whales of the Bight is still painful. It was, he says, 'like watching our children being killed'.[89] This deep connection with the southern right whales of the Bight inspired a remarkable sequence of events that resonated around the world. Lawrie heard about a British actor turned filmmaker, Kim Kindersley, who had been inspired to produce documentaries about interactions between wild dolphins and humans after a deeply moving experience while diving off the west coast of Ireland, when he recognised the human capacity for spiritual connection with the marine mammal.[90] Lawrie invited Kindersley to witness a traditional whale-calling ceremony at the Head of Bight, in the Southern Ocean. The two met and, as Lawrie sang and Kindersley filmed, a southern right whale leapt from the ocean and roared.

Kindersley's footage became the catalyst for a gathering of indigenous leaders whose people had spiritual ties to dolphins and whales. From other continents and other oceans, indigenous elders converged on the Head of Bight to sing in the whales and share their peoples' stories. Amongst them was Alberta Thompson, from the Makah, a Pacific tribe from northwestern America, who attended to protest the slaughter of the grey whale on the shores of her traditional country; and Terry Freitas, who travelled from the cloud forest of the Amazon River, in South America, to share his U'wa tribe's song about the interconnectedness of humans and all creatures. The resulting film, *Whaledreamers*, was released for public viewing in 2008 and won eight international awards. It offered a powerful message about human connectedness with the natural world.[91]

The event was also a powerful testimony to the endurance of the Mirning people, who had been forcibly removed from their traditional country to clear the way for European colonisation in the nineteenth century. During the 1950s, in the midst of the Cold War, the British Government had decided that Maralinga and Emu Field, in the remote desert country north of the Nullarbor, were suitably remote for the testing of nuclear weapons.[92] For its part, the Australian Government relocated the Aboriginal inhabitants of Maralinga to the Head of Bight and granted them a 99-year lease over the Mirnings' ancestral country. So began a long legal battle by the 'people of the whale' for recognition of their native title of land and sea. On 24 October 2017, after a 22-year legal process, the Federal Court of Australia formally recognised native title for the Mirning people, covering about 33 000 square kilometres, including two-thirds deemed to be the 'exclusive possession' of the native title holders. This determination was recognised in a ceremony held at a site near Mundrabilla, on the Great Australian Bight.[93]

Meanwhile, the whales continued to follow their ancestral pathways to the Bight. In 1979 the Australian Government introduced a ban on commercial whaling, and in 1998 it declared the Bight to be part of a marine park, embracing the waters of the Southern Ocean from Nuyts Reef, in southwestern Western Australia, to the Bunda Cliffs. The park recognised the Bight as a globally significant nursery for the threatened southern right whale and an important feeding habitat for the Australian sea lion, white shark, migratory sperm whale and short-tailed shearwater.[94] After nearly a century of hunting, it seemed that the southern right whale would finally find sanctuary in the Southern Ocean, below the Bunda Cliffs. But the whales that migrate each year between their Antarctic feeding grounds and the Bight live on borrowed time. In August 2016 Bob Brown, a conservationist and former Australian Greens senator, boarded Sea Shepherd Australia's vessel *Steve Irwin* to add his support to a campaign to stop the installation of an oil and gas rig in the Great Australian Bight Marine Park. The voyage aimed to draw public attention to the potentially catastrophic consequences for the Bight's fragile marine ecosystem should oil and water converge in this most volatile of oceans.[95] The issue was the subject of a short-lived parliamentary inquiry. The British oil and gas company involved withdrew from the Bight but another company, from Norway, announced its intention to search there for oil and gas. As southern right whales continue to rally in the sheltered waters of the Bunda Cliffs, another threat to their survival stirs.[96]

7
CONVERGENCE

We cannot think of a time that is oceanless
Or of an ocean not littered with wastage
Or of a future that is not liable
Like the past, to have no destination.

TS Eliot, 'The dry salvages', 1941[1]

Between Latitude 54° 38′ South, Longitude 35° 47′ West and 61° 05′ South, 54° 52′ West: between Gold Harbour, South Georgia, and Point Wild, Elephant Island (4-7 November 2017)

After days of slow swells, gale-force winds finally reach us from the other side of the planet. The entry into Drygalski Fjord, at the southeastern end of South Georgia, is a cauldron, and the ocean throws everything at us as we round Cape Disappointment at the southern extremity of the island. This force-10 storm reminds us that nature has the last word in this ocean. James Cook named the cape in 1775 when he realised that he had charted an island rather than the southern continent. Far from being disappointed, I am transfixed by the colours of the sky. A luminous sun floods the ragged white peaks, and strange golden hues are interspersed with bright blue and deep iron clouds.

Once the ship starts to roll, we finally understand the true meaning of 'one hand for the ship'. Negotiating a passageway or stairs with both hands full is simply not an option, and all loose items are stowed, or else they become small missiles that whiz across the cabin. Buffeted by storm-force winds, I seek refuge in the ship's library as ice coats the windows, and read Herbert Ponting's description of the Furious Fifties, written in his cabin aboard the *Terra Nova* in 1910. The former whaling vessel was in the midst of such a storm just three days south from New Zealand. Ponting had been looking forward to a true Southern Ocean gale in the 'stormiest zone on earth', but when it came, a leg injury prevented him from climbing to the bridge to witness its grandeur. He wrote,

> The ship rolled and plunged and squirmed as she wallowed in the tremendous seas which boomed and crashed all that night against the weather side sending tons of water aboard every minute. Screaming gusts would strike her with hurricane force, and sometimes she would lay over to an angle of 40° – nearly half a right angle from an even keel.[2]

I cannot help but contemplate Ernest Shackleton and Frank Worsley's extraordinary 1500-kilometre voyage through this region, when they and four others sailed from Elephant Island to South Georgia in a cramped lifeboat so that

22 marooned men might be rescued from a narrow spit of rock and ice as winter closed in. As I peer into the nebulous horizon of sea and fog, I am quietly in awe of Worsley's navigational skills.

In the first edition of *The Sea Around Us*, published in 1951, the American marine biologist Rachel Carson wrote that the world's oceans were 'inviolate, beyond man's ability to change and to despoil'.[3] By the time of the second edition, in 1960, the international scientific community was divided over the state of the world's oceans. Some saw them as an infinite resource akin to the emerging frontier of outer space. The deep ocean seemed to be the last great unexplored region on Earth, merely awaiting human technology to discover its riches and bring its fruits to bear. The American science journalist Robert Cowen enthused in his book *Frontiers of the Sea*, also published in 1960, that the ocean's 'prospects [shone] with the promise of a virtually inexhaustible supply of minerals and metals, and of a significant increase in the world's food supply'.[4] Others advocated more research before such actions were taken, questioning the capacity of the world's oceans to absorb the impact of increasing human exploitation and pollution. The problem for ocean scientists was how to measure the impact of human activities on the marine environment and provide the appropriate guidance to policymakers.

The study of the geography and physical nature of oceans became increasingly specialised in the post-war era, with new national programs seeking to shed light on some of the more elusive aspects of the deep-ocean environment. The Swedish geographer

Hans Pettersson, for example, led the first detailed geological study of the deep-ocean floor in 1947–8, aboard the training ship *Albatross*, using a newly developed coring device that could penetrate the seabed up to 20 metres, and echo sounders that could produce profiles of the seabed in waters up to 7000 metres deep.[5] As we saw earlier, soon afterwards Denmark also became involved in deep-sea exploration with the *Galathea* expedition. Britain continued its oceanographic research in the Southern Ocean too. It had been a prominent contributor to scientific programs in the decades before World War II and, while that war brought a temporary halt to its work, *Discovery II* returned to the Southern Ocean in 1950 for its final year of research under the auspices of the new National Institute of Oceanography of Great Britain. Its mission was to map currents and sea temperatures at different levels and to examine the nature of the ocean floor and plot the biological and chemical characters of its waters. The renowned oceanographer George Deacon was on its scientific team. Beginning in 1927 as a chemist aboard the *William Scoresby*, he had subsequently transferred to *Discovery II*, tracking the movements of the Southern Ocean's water masses between the Antarctic continent and the Antarctic Convergence. By the time the Discovery Investigations came to an end, the three vessels involved had collectively made 14 expeditions into the Southern Ocean, the results of which were published in 38 volumes. The British-born Australian zoologist William Dakin declared it to be 'one of the most successful efforts made by science to solve some of the problems of the sea'. He noted that, while the project had yielded valuable information about the behaviour and breeding of whales in the region, much more research was needed to understand how ocean circulation supported marine life. 'Once upon a time', he wrote in 1947, 'explorers visited the Arctic and

Antarctic Seas for the excitement and for the honour and glory of discovering new land – now it is the turn of the scientist. Let us hope to the advantage and peace of mankind.'[6]

His words seemed prophetic. Within a decade 65 nations had agreed to participate in the International Geophysical Year, a significant achievement given the politically charged climate that characterised East–West relations during the Cold War. Projects in the Antarctic region involving some 30 000 scientists focused on the physical properties of Earth and its atmosphere, meteorology, seismology, geomagnetism, and ionospheric and auroral physics. The scientific importance of the region was recognised, with 12 nations (Argentina, Australia, Belgium, Chile, France, Great Britain, Japan, New Zealand, Norway, South Africa, the United States and the Soviet Union) agreeing to establish scientific bases giving scientists access to previously unexplored areas of the continent. Conducted between July 1957 and December 1958, it heralded a new era of international scientific exploration of the region, as well as of the floors of the world's oceans and outer space. It was hailed as the most ambitious collaborative scientific project ever conceived and a remarkable demonstration of how nations could work together.

The International Geophysical Year's success also provided the impetus for the formation of the *Antarctic Treaty*. Initially signed by 12 nations in 1959, the treaty set out governance arrangements for the continent, islands, iceshelf and ocean south of latitude 60° South and encouraged collaborative scientific research and exchange of information, with the Scientific Committee on Antarctic Research co-ordinating research activities. In the words of the treaty, Antarctica was 'forever to be used exclusively for peaceful purposes and [should] not become the scene or object of international discord'. The treaty was regarded as a landmark in international politics, and in 1991 it was reaffirmed and said to be one of the most successful international agreements of its kind.[7]

Oceanography was the first of the burgeoning sciences to hold an international congress, and 800 delegates from 38 countries met in 1959 at the United Nations in New York to review the current state of knowledge about the history and nature of the world's oceans.[8] The United States expanded its research program in the Antarctic region by launching one of the most significant national oceanographic research programs of the post-war era. Between 1962 and 1975 the US naval ship *Eltanin* made 55 expeditions, collecting data along 777 840 kilometres of track between latitude 40° South and the Antarctic continent. Sayed El-Sayed, one of the senior scientists involved in the program, noted that the simplicity and instability of the Southern Ocean made it particularly vulnerable to human activities and the impacts of these, he believed, could be properly assessed only by investigating the ocean environment as a whole rather than by focusing on the biology and behaviour of individual species as the Discovery Investigations had done.[9] By 1971, as the Australian marine biologist Isobel Bennett noted, research vessels from Japan, the Soviet Union and New Zealand, as well as from the United States, were actively exploring the waters of the Southern Ocean, employing new devices to trawl its depths and electronic flash underwater photography to reveal life under the sea ice.[10]

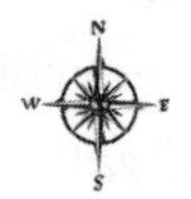

The oceans and seas of the world have long been the focus of conflict between competing maritime nations seeking to assert their authority over particular tracts of water and resources. The question of who owns the right to strategic shipping routes dates back to at least the second century CE, when Roman law decreed the sea to be *res communis* (the property of all).[11] This was enshrined in *Mare liberum* (*The Freedom of the Seas*), written in 1609 by the Dutch jurist and philosopher Hugo Grotius, in defence of the Dutch right to trade in South Asia against Spanish and Portuguese claims to exclusive use of newly discovered ocean routes. It assumed great significance as European maritime empires began to extend their reach into previously unknown oceans and as competing mercantile states sought to dominate strategic shipping routes.[12]

By the early nineteenth century the oceans were perceived not only as a means to extend territorial ambitions but also as the source of inexhaustible riches unencumbered by any one nation's legal or administrative instruments. It was the principle of the freedom of the seas that effectively sanctioned indiscriminate sealing activities in the Southern Ocean, with sealers becoming the advance guard for the wholesale exploitation of marine mammals and birds in countless remote bays and beaches of the subantarctic and Antarctic region.[13]

A scientific review of whale sightings around South Georgia between 1979 and 1998 noted that humpbacks had been commercially extinct by 1915, Antarctic blue whales by 1935 and fin and sei whales by 1960. It concluded that all of the southern whale species that once frequented the waters had become rare as a result of the southern whaling industry. Southern right whales were classed as commercially extinct even before the modern era of shore-based whaling began in 1900, and became the first whale species to be protected globally by the International Whaling Commission, from 1932. Blue whales were thought to be in particularly dire straits. According to estimates by whale researchers, commercial whaling had reduced the global population from around 300 000 to just 6000 individuals.[14] A yachtsman who sailed regularly in the waters around South Georgia after the cessation of pelagic whaling in 1965 reported seeing only one female blue whale and her calf over a 25-year period. In 2014, researchers estimated that 2 053 956 whales of all species were slaughtered in the Southern Ocean between 1900 and 1999, with a further 563 696 in the North Pacific Ocean and 276 442 in the North Atlantic Ocean. [15]

Perhaps it was fitting that the near extermination of the Southern Ocean's iconic whales should become the catalyst for an environmental campaign that brought that remotest of oceans into people's living rooms around the world. In 1970 the environment activist organisation Greenpeace International launched its first public anti-whaling campaign, against Soviet whaling fleets in the Pacific Ocean. Using graphic images of modern whale-hunting practices and controversial 'direct action' tactics, navigating inflatable boats between whalers and whales, the Save the Whales campaign heightened public awareness of the brutal realities of whale hunting and the extent of industrialised whaling at a time when whale products were no longer in demand. By the 1960s, for example, margarine manufacturers were using polyunsaturated fats from vegetable sources in preference to whale oil, and petroleum had long replaced whale oil as a source of fuel.[16]

In Australia an environmental group founded in New Zealand called Project Jonah led an aggressive campaign to pressure the government to close its last whaling station at Cheynes Beach near Albany in Western Australia. The station had begun its operations in 1952 with a licence to process 50 humpback whales per year and at its peak was processing over 1000 sperm and humpback whales annually. A ban on humpback whaling introduced in 1963 forced the company to concentrate on the sperm whales, although finding buyers for sperm oil became increasingly difficult. By 1977 anti-whaling sentiment was reaching fever pitch in the old whaling town of Albany, where protesters led by French activist Jean-Paul Fortom-Gouin and calling themselves the Whale and Dolphin Coalition steered rubber boats into the path of whaling vessels and their harpoons. The confrontations became intensely personal for whalers and protesters alike, and feelings ran high.[17]

Ultimately, the combination of declining markets and anti-whaling sentiment brought Australia's whaling industry to a close, and Cheynes Beach ceased operating in 1978. In the following year the Australian Government introduced a ban on commercial whaling within the country's economic exclusion zone, and two decades later Australia declared the waters around the continent and its external territories (including Macquarie, Heard and McDonald Islands, in the Southern Ocean) a sanctuary for the protection of whales and dolphins.[18] Meanwhile, mounting government and public opposition to commercial whaling in many nations persuaded the International Whaling Commission to declare the Indian Ocean Whale Sanctuary in 1979, followed by an international moratorium on all commercial whaling, in 1982.[19]

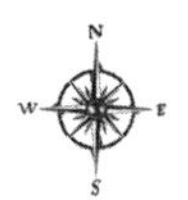

In the years after World War II, many traditional fishing grounds in the Northern Hemisphere faced catastrophic collapse as a result of over fishing, including the Californian sardine fishery, in the 1950s; the Scandinavian herring fishery, in the late 1960s; the

Peruvian anchovy fishery, in the early 1970s; and the northern cod fishery in the Atlantic Ocean, in the early 1990s. Fishing enterprises cast their gaze towards the high seas of the Southern Ocean as the last frontier for large-scale commercial fisheries.[20] The *Antarctic Treaty*, signed in 1959, had created an international framework for managing human activities on the Antarctic continent – those affecting land, ice and water south of latitude 60° South – but it did not provide any protection for the living marine resources of the Southern Ocean north of that latitude and a most unlikely species became the focus of another phase of exploitation.[21]

Whalers had long known of the importance of Antarctic krill as the principal source of food for baleen whales, and they were alert to the distinctive reddish stain in surface waters that indicated the presence of a krill swarm. Antarctic krill are small shrimp-like crustaceans that grow to about the size of a human thumb, and they are one of the most abundant animal species on Earth. One estimate puts the krill population in the Southern Ocean at two quadrillion individual creatures.[22] Antarctic krill breed south of the Antarctic Convergence – between latitudes 48° South and 62° South – but they are not found any further north. Indeed, the presence (or absence) of krill is a sure sign that voyagers have crossed the convergence, such is the nature of this biological frontier.[23] Krill have prominent black eyes and transparent bodies, but they sport a bright red pigmentation on their shells, which gives them a distinctive pinky-orange appearance. Individual krill can live for up to six years, and their life cycle is both remarkable and curious. Adult females spawn in summer; their eggs drift down into the ocean depths, where they hatch in darkness. The larvae then have ten days or so to rise to the surface, where they feed on phytoplankton. They take three years to mature, and some are carried great distances by the strong currents around Antarctica.

Krill are remarkable for being constantly on the move, fluttering all five pairs of feathery legs at once to propel themselves through the water and to prevent themselves from sinking. The effect is mesmerising, as Francis Downes Ommanney observed in 1938:

> The 'krill' is a creature of delicate and feathery beauty, reddish brown and glassily transparent. It swims with that curiously intent purposefulness peculiar to shrimps, all its feelers alert for a touch, tremulously sensitive, its protruding black eyes set forward like lamps. It moves forward slowly, deliberately, with its feathery limbs working in rhythm and, at a touch of its feelers, shoots backwards with stupefying rapidity to begin its cautious forward progress once again.[24]

Krill also have a distinctive habit of forming swarms, which can cover many square kilometres of surface water and reach many metres into the depths, their immense numbers turning the sea red. Wherever krill swarm, there is every chance that their predators will also appear. They are the favoured food of baleen whales, dolphins, seals, penguins and seabirds. Alister Hardy, the chief zoologist with the Discovery Investigations, admired the beauty of krill and found pleasure in painting individual creatures in watercolours. 'For a whole day', he recalled in 1956, 'there was a dense swarm, like a red cloud, of closely packed Euphausians (*Euphausia superba*) against the jetty at our shore station … The cloud would sometimes change shape, elongate this way or that.' To Hardy, the krill's movement seemed almost to be made in response to an invisible command from their leader.[25] More recently, John Kernan, an American marine zoologist and Antarctic diver, found himself in the midst of a krill swarm during an expedition into Charcot Bay, on the northernmost part of the Antarctic Peninsula. He was surprised to see large numbers of penguins, seals and humpback whales in the bay but then realised that his boat propeller was spitting out krill onto the ice. The swarm spread out for kilometres across the sea, and when the expedition's divers filmed it under the surface, they found that it also extended well below 30 metres.[26]

In 1962 the Scottish zoologist and marine biologist James Marr published a landmark study of Antarctic krill, drawing on three decades of observations and data gathered during the Discovery Investigations and providing valuable information about their location and behaviour. Marr had participated in the investigations and in 1949 had become the principal scientific officer at the National Institute of Oceanography, in Britain.[27] Amidst speculation that the over-exploitation of whales in the Southern Ocean had resulted in an abundance of Antarctic krill, and that krill could provide a valuable new source of protein for humans, the Soviet Union sent a fisheries research vessel into the Southern Ocean in 1964 to assess krill stocks and to determine whether they could be harvested on a commercial basis.[28] A new industry was born, and by 1970 the Soviets were marketing krill butter and krill cheese, as well as a red-coloured krill paste which was described as having nutritional value and a 'delicate, sweetish flavour typical of shrimp'. While krill was not to everyone's taste, the Soviets enthusiastically advertised krill paste as an ingredient to enrich the flavour of dishes such as Siberian dumplings and promoted it as a cure for stomach ulcers and other common human ailments. The paste residue was also marketed as a nutritious addition to animal feed. Processing fresh Antarctic krill to make it suitable for human consumption posed certain logistical challenges in the remote Southern Ocean. Apart from krill swarms being difficult to locate in the vast waters, the krill needed to be cooked or frozen within two hours of being caught in order to prevent deterioration. Before long scientists were warning that the scale of the harvesting was creating problems for the southern baleen whales. In 1975 the *New York Times* reported on the rise of the Antarctic krill industry, arguing that it posed 'new ecological problems' and citing concerns raised by marine scientists that large-scale krill harvesting was having an adverse impact on the larger marine mammals that depended on them.[29] By this time, the Southern Ocean krill had become firmly entangled in the political machinations of the Cold War.[30]

At the eighth consultative meeting of *Antarctic Treaty* members in 1975, options were discussed for a legal agreement that would conserve the living marine resources of the Southern Ocean. Later that year, members of the Scientific Committee on Antarctic Research established the scientific program Biological Investigations of Marine Antarctic Systems and Stocks (BIOMASS) to investigate the 'living resources' of the Antarctic marine ecosystem and outline the measures necessary to manage their numbers in a sustainable way.[31] The British ecologist Arthur Tansley had first used the term 'ecosystem' in 1935, to describe how living organisms and the physical environment form a holistic and integrated system.[32] The word became more widely known with the publication in 1953 of Eugene Odum's *Fundamentals of Ecology* and the application of ecosystem principles gained momentum with the introduction of land-based environmental protection programs, culminating in the United Nations Educational, Scientific and Cultural Organization's Man and Biosphere program, in 1971.[33] This paved the way for a reimagining of the Southern Ocean, not simply as a habitat for individual species, but as an interconnected whole in which changes affecting one species could be expected to impact on other species, as well as on their physical environment. When Richard Laws, director of the British Antarctic Survey, addressed a special consultative meeting of Treaty Parties in 1978 on the issue of conserving Antarctic marine living resources, he described the Southern Ocean as an ecosystem which, he explained, comprised 'a volume of ocean with unique physical and chemical properties and all the living organisms within it, the structure of the communities they form[ed], the dynamic functions and the biomass of the organisms and different trophic levels, and the complex interactions of species with each other and the environment'.[34]

It was shaping up to be a momentous decade for the world's oceans and seas. In 1982 agreement was reached on the *United Nations Convention of the Law of the Sea*, a comprehensive legal regime that replaced the traditional concept of the freedom of the seas with recognition of national sovereignty and economic goals alongside promotion of peaceful uses of the world's oceans and seas. It accepted customary international law regarding territorial rights to coastal resources and confirmed that the world's oceans were the common heritage of humankind. In particular, the convention acknowledged the inherent problems of managing something as amorphous as 'ocean space'.[35]

In the same year, the *Convention on the Conservation of Antarctic Marine Living Resources* (CAMLR Convention) came into force. It was signed by the 12 original signatories to the *Antarctic Treaty* amidst international concern about the rapid expansion of the Antarctic krill fisheries by the Soviet Union and Ukraine (and to a lesser extent by Japan).[36] The aim of the Convention was to regulate the harvesting of species in what was regarded as one of the world's last remaining wildernesses. It provided for the establishment of an international commission which recognised that some nations would continue to exploit the ocean's resources, but declared that they should not deplete its 'stocks' or adversely affect other species in the ecosystem. Member nations, which numbered 25 by 2018, agreed to pay a subscription based on the amount of resources that they took each year.[37]

The Convention extended the reach of international governance under the *Antarctic Treaty* to include all marine living resources as far north as the Antarctic Convergence, around latitude 55° South, and recognised that predators and prey formed part of an interconnected marine ecosystem. It recognised that Antarctic krill was the keystone species of the Southern Ocean ecosystem and that unregulated krill fishing – or simply a poor breeding season – could have a detrimental impact on all species that depended on it for food, and especially on those with no alternatives.[38] Off South Georgia, for example, the numbers of some marine creatures had declined as a direct result of the disappearance of krill. The Convention also required member nations to take the needs of krill predators into account when making decisions in relation to harvesting, and to exploit the ocean's living marine resources in such a way as to prevent irreversible changes to the rest of the ecosystem. The commission established agreement on which species were to be protected from exploitation, catch limits, where and how fishing would be undertaken and prosecution processes for violations.[39] In addition to Antarctic krill, the commission aimed to regulate the killing of finfish, molluscs, crustaceans and seabirds. It specifically excluded the conservation of whales and seals, which were dealt with under the 1946 *International Convention for the Regulation of Whaling* and the 1972 *Convention for the Conservation of Antarctic Seals* respectively.[40] The Soviet krill industry peaked in the 1980s, but it never instigated the 'gold rush' anticipated by the CAMLR Convention and Antarctic krill continued to be harvested, under licence, for use in industries such as aquaculture and as a source of omega-3 fatty acids for human consumption. Nevertheless, the Convention was credited with having effectively regulated the industry and enshrining the concept of ecosystem conservation in international law.[41]

Adding to the urgency of this regulation had been concern since the 1970s about an unexplained long-term decline in the populations of adult Antarctic krill in the Southern Ocean, particularly in the waters between the Weddell Sea and the Antarctic Peninsula, where by 2016 numbers had declined by an estimated 70 to 80 per cent over the previous 40 years.[42] Even by 2018 no agreement had been reached on whether the decline was in response to climate change, to the recovery of whale numbers after the end of commercial whaling or to a combination of factors. Researchers from Yale University's Institute for Biospheric Studies issued a warning in 2016, for example, about how predicted rises in ocean temperatures and reductions in sea ice around Antarctica during the course of the twenty-first century were likely to affect the young krill that congregated under the ice to feed: the outlook for one of the most important species of the Southern Ocean's ecosystem seemed grim.[43]

With the success of the CAMLR Convention, it seemed that managing the ocean as an ecosystem might address the 'boom and bust' exploitation pattern characteristic of humankind's treatment of individual marine species of the Southern Ocean since the start of the sealing industry in the late eighteenth century.[44] Nevertheless, the question of who should have access to the ocean's resources continued to be a source of international disputes.[45] In 2002, for example, 21 scientists published an open letter to the *New York Times* critical of the methods used by Japan in its scientific whaling program and questioning its motives.[46] Japan, which had a long history of whale hunting, began its program in the Southern Ocean in 2002 under the provisions for whaling research in the International Whaling Commission's moratorium. In 2014, however, following a concerted campaign by the Australian Government to regulate activities within Australia's whale sanctuary, the International Court of Justice ruled that Japan's 'scientific' whaling program in the Southern Ocean as a whole contravened the *International Convention for the Regulation of Whaling* and determined that it was not for scientific purposes as claimed. Despite this, the legal battles continue in Australian courts over Japan's whaling activities in the Southern Ocean.[47]

Other challenges to the Southern Ocean's ecosystem management framework arose by the turn of the twenty-first century. In 1991 member nations of the *Antarctic Treaty* agreed to the *Protocol on Environmental Protection to the Antarctic Treaty* (also known as the Madrid Protocol) in an effort to strengthen the treaty system and regulate mining activities in Antarctica.[48] Meanwhile, the rise of the lucrative industry of commercial fishing for the highly prized Antarctic toothfish and Patagonian toothfish, known as 'white gold', was attracting illegal and unregulated operations that used longlines to trawl the high seas of the Southern Hemisphere. In December 2017, after ten years of discussion, member countries of the United Nations agreed to negotiate a conservation treaty to regulate illegal fishing on the high seas beyond the exclusive economic zones of countries with coastlines. However, even the legal use of longline hooks has taken a heavy toll on the ocean's marine life, with many thousands of albatrosses and other seabirds drowning as they dive and swallow the baited hooks.[49]

Studies of the ceaseless movements of the Southern Ocean and its powerful alliance with Earth's other natural forces continue. The American geoscientist Wallace 'Wally' Broecker revolutionised oceanography in the 1980s, when he discovered the link between ocean circulation and climate change and coined the term 'great ocean conveyor' to explain how sudden changes in the circulation of deep-ocean currents triggered past ice ages. As a result, research efforts intensified to discover the role of the Southern Ocean in regulating global climate patterns.[50] Robotic Argo floats (named Argo after a 1999 scientific program) were developed, capable of transmitting data on temperature, salinity and currents all year round, including from beneath sea ice to distant vessels via satellites orbiting in space. The Integrated Marine Observing System began operating such equipment in Australian waters in 2006. During the fourth International Polar Year, in 2007–8, the International Council for Science, the World Meteorological Organization, the Arctic Council, members of the *Antarctic Treaty* and other international organisations combined forces to examine the influence of the two polar regions on climate.[51] In the following year, Australian and American scientists involved in a joint research program found that winds blowing on the Southern Ocean were strengthening and creating changes in the surface layer of the water, with implications for the way in which the ocean and atmosphere exchange heat and carbon dioxide.[52] In 2012 an international team of oceanographers announced the creation of the Southern Ocean Observing System, to remotely plot the circulation of the ocean's deep currents, and produce data that would help to explain their role in shaping global climate patterns and cycling carbon and nutrients.[53] By deploying underwater autonomous vehicles and CTD profilers to continue the time-honoured tasks of measuring depth, temperature and salinity, ocean scientists were capable of taking the pulse of the ocean from the deepest Antarctic Bottom Water to the surface.[54] They also began to track the global ocean circulation system, mapping changes in the physical, biological and biogeochemical properties of the ocean beneath sea ice, and to demonstrate how deep water from the sea floor is in touch with the atmosphere, regulating global atmospheric temperatures. Through technology, scientists are now visualising what, in 1757 Sir Benjamin Thompson, Count von Rumford, was only able to theorise: that the atmosphere and ocean are deeply interconnected.[55]

The scientific crusade in the high southern latitudes by early in the twenty-first century was no longer just a matter of curiosity or economic gain. Pressured by mounting evidence of anthropogenic climate change, yet conflicted by moral and political uncertainty, those studying the changes occurring in the Southern Ocean felt a growing sense of urgency. The Southern Ocean was no longer considered simply a remote, sublime space devoid of human habitation; indeed, their research had proved that Earth is profoundly dependent upon the ocean's heartbeat of seasonal ice, its carbon-filled lungs and the slow circulation of its deep currents. Oceanographers now saw the Southern Ocean as a massive global engine, storing heat, moisture and carbon dioxide and transferring them thousands of kilometres, from one part of Earth to another, shaping temperature and rainfall patterns far to the north.[56]

According to ocean scientists the Southern Ocean, for so long shrouded in legend, had become a 'window to the deep sea', but what it revealed was disturbing.[57] Those studying changes in the temperature and salinity of the Southern Ocean focused on the implications of such changes for the global ocean. The ocean's microscopic diatoms and other phytoplankton that inhabit the briny veins of sea ice and stain it yellow not only serve as food for Antarctic krill; they also carry out almost half the photosynthesis on Earth, absorbing carbon dioxide into their delicate skeletons. When they die, the skeletons float down to the ocean floor, taking the carbon dioxide with them into the ooze and locking it up for thousands of years.[58] In 2012 oceanographers discovered that the normally dense bottom water of the Southern Ocean was considerably less dense than it was in the 1970s, and this was reducing the capacity of the ocean to store carbon dioxide.[59] While all the world's oceans have the capacity to absorb heat and carbon dioxide, the Southern Ocean absorbs more than any other.[60] Scientists recognised that the Southern Ocean, like the Antarctic continent itself, is a barometer of global warming.[61] The Australian physical oceanographers Nathan Bindoff and Steve Rintoul and political scientist Marcus Haward were not alone amongst Southern Ocean experts when they expressed the view in 2011 that 'Southern Ocean processes are intimately linked to some of the most pressing challenges faced by society: climate change, sea-level rise, ocean acidification and the sustainable management of marine resources'.[62]

When the Australian oceanographer and climate change scientist, Matthew England, appeared before a hearing of the

Australian parliamentary Joint Standing Committee on Treaties in 2016, his testimony went to the very heart of climate science and the implications of global warming for the driest inhabited continent on Earth.[63] One of the most pressing problems in the Southern Ocean at that point was increasing acidification – a chemical change caused by the build-up of greenhouse gases in Earth's atmosphere. Ocean acidification would have catastrophic impacts on marine species with shells made from calcium carbonate, as it would reduce the amount of carbonate available to make their shells and lessen their ability to transfer carbon into the depths and release oxygen into the air.[64] In 2016, a team of marine biologists aboard the British Antarctic Survey vessel the *James Clark Ross* began examining communities of benthic marine life along the continental shelves of the Southern Ocean. The Antarctic Seabed Carbon Capture Change project aimed to discover the role of the ocean in cycling carbon, and how much carbon it takes with it to its final resting place on the seabed. Its scientists were learning to read deep time in the deep ocean.[65]

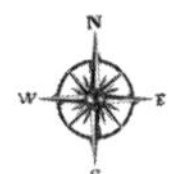

Within a century, scientific perceptions of the Southern Ocean have shifted dramatically, from an abundant resource ripe for exploitation to a vulnerable ecosystem in which the whole of humanity – not just individual nations – has a vested interest.[66] For all these efforts, however, the Southern Ocean still seems wild and isolated, far removed from the lives and concerns of most people. We find it difficult to comprehend its influence on our lives or our impact on its well-being, because we do not live there, in that place. Some occupy its edges and traverse its surfaces but, even as it drives the planetary forces that support life on Earth, it seems peripheral to human affairs. We might study an individual species in order to understand its behaviour, to exploit it or to manage it, but it seems that the Southern Ocean environment as a whole will forever be something of an enigma. If land grounds us and imparts a sense of place and identity, ocean confounds us with its fluidity and restlessness and inaccessibility. How can we truly know a place that we cannot inhabit and that most can see only via the medium of technology? Why should we care about the Southern Ocean, where few people live and which few experience first hand? Is it possible to develop an ocean consciousness? These are some of the questions that have framed the research for this book.

Oceans challenge us to think differently about our planet. Everything is interconnected and fluid, with masses of water moving around the globe on a scale that defies the imagination. Our inability to experience first hand the three-dimensional ocean environment – mostly beyond the reach of any light – profoundly shapes how we think about it. Such an environment refuses to succumb to the neat lines of the cartographer or hydrographer or policymaker. Even at the surface, the Southern Ocean's immensity and wildness are overwhelming. We need to use our imagination to engage with this environment, but how can we start to imagine a place that is so inaccessible and volatile?

Bringing the Southern Ocean's natural and cultural histories into the light allows us to see it as more than a remote scientific laboratory or as a source of food or prosperity. In rendering the Southern Ocean as a space for industry and science, we have become curiously disengaged from this wild, little-known sea. Classic sea stories such as *Moby Dick* and *Twenty Thousand Leagues*

Under the Sea were written in the nineteenth century, during the age of sail and long-distance ocean voyaging.[67] But where are the ocean stories for the twenty-first century? More particularly, where are the Southern Ocean stories? Modern natural history films like the BBC's *Blue Planet* documentary series have done much to foster a popular 'sense of wonder' in the world's oceans, to borrow Rachel Carson's phrase.[68] But where is the literature that inspires us and perhaps also invites us to consider the ocean's fate in the Anthropocene?[69]

In 2016 the Indian writer Amitav Ghosh challenged his readers to consider why literary fiction had failed to engage with climate change. 'Is it perhaps too wild a stream to be navigated in the accustomed barques of narration?' he asked.

> The truth, as is now widely acknowledged, is that we have entered a time when the wild has become the norm: if certain literary forms are unable to negotiate these torrents, then they will have failed – and their failures will have to be counted as an aspect of the broader imaginative and cultural failure that lies at the heart of the climate crisis.[70]

The stories in this book offer another way of seeing the Southern Ocean. They reveal how the indigenous peoples of the coasts and islands of the Southern Ocean have intimate relationships with their sea country and its creatures, and how microscopic diatoms and translucent krill are as much part of the ocean's history as trawl nets and try-pots. Krill may be the fluttering heart of the ocean's vast food chain, for example, but the humble creatures are invisible in popular perceptions of the Southern Ocean and its marine life. As the Tasmanian literary scholar Elizabeth Leane and krill biologist Steve Nicol have written, even scientists are inclined to dismiss krill as inanimate fodder for larger marine species, rather than recognising their miraculous presence in the ocean and seeking to understand their social behaviour and interactions with their environment.[71] The stories here show that, far from being a wild sea at the uttermost end of Earth, the Southern Ocean is deeply entangled with humanity's past and the world's future; it demands our close attention. Tony Press, a former director of the Australian Antarctic Division, wrote in 2002 that the Southern Ocean 'touches the lives of us all … a living, breathing, organic body of water that embraces us and nurtures us in ways we cannot yet fully appreciate'.[72]

Perhaps the greatest gift of the Southern Ocean is a sense of connection: catching a leopard seal's eye, marvelling at a swarm of krill, hearing the triumphant cacophony of king penguins in breeding season, absorbing the unearthly blue of an iceberg, admiring the majesty of the blue whale and being left speechless by the extraordinary power and pureness of purpose in the flight of the wandering albatross. This book has distilled the Southern Ocean to its core elements. It may not be possible to feel a sense of belonging in this vast, complex, fluid environment, but it seems that we can have a sense of connection – with ocean and wind and ice and the myriad of living things. Ultimately, we are all caught within the Southern Ocean's embrace.

Latitude 55° 05′ South, Longitude 66° 47′ West: crossing the Drake Passage (11 November 2017)

Apart from grey swells and windswept clouds, there is little to catch the eye on the surface of this most notorious ocean passage. The light of the Southern Ocean plays tricks with the eyes, obscuring the horizon so that sea and sky meld. Sometimes it seems that we are no longer voyaging between sea and sky but have become enshrouded by both. The pictures I have held in my mind come to the surface now, inhabiting the great expanses of the high southern latitudes. In the not-quite-darkness of these latitudes in summer, I wonder if this is what the early sailors meant by 'looming' – seeing things beyond the limits of normal vision. I consider the krill swarming around us as they come to the surface to feed. Facing westward, I sense the powerful currents pummelling against our hull. I reflect on the stillness in the ice, the lulling rock of the ship, my admiration for and awe of the migration of birds and whales, the clarity of thought in the polar air.

As a historian interested in sense of place and identity, I have been compelled to reflect on my own encounters with the Southern Ocean. Given the difficulties of experiencing its vast expanses in any cohesive way, I have drawn on journeys undertaken over several years to different places and latitudes. I have kept company with the Roaring Forties during trips to remote vantage points in southern Australia, New Zealand and South America. I have sailed southward into the Furious Fifties and Screaming Sixties to reach the remote beaches of South Georgia, the South Shetlands and the Antarctic Peninsula, and I have crouched low in small boats, taking shelter from the waves and winds, so that I could witness the ethereal beauty of the ice. Much remains to be seen but this place, this ocean, has changed me forever.

Cape Horn comes into view – a ragged black promontory that holds so much human misery close to its diminutive chest. I see the albatross keeping its lonely vigil at the end of Earth, a memorial created by the Chilean sculptor José Balcells Eyquem for all those who have drowned off the Horn. 'But they did not die in the furious waves, today they fly on my wings, toward eternity, in the last crevasse of the Antarctic winds.'[73] The beautiful pintado petrels escort our ship as we round the cape, flashing intricate black-and-white painted wings. The small birds have been with us since South Georgia, conducting their aerial acrobatics around the vessel and seeming to never quite touch the water.

ACKNOWLEDGMENTS

I would like to thank Tom Griffiths for his friendship and inspiration over the years, and for his generosity in involving me in the Centre for Environmental History at the Australian National University. I am also grateful to Tom and Libby Robin for hosting many wonderful gatherings of environmental historians from around the globe at their home in Canberra. Narissa Bax, Mike Coffin, Kirsty Douglas, Tom Griffiths and Bernadette Hince have read draft chapters of this book, and I thank them for giving so generously of their time and expertise to provide insightful comments and constructive suggestions for improvement.

The ANU School of History has given me much valuable support over the past three years by providing an office and enabling me to be part of a supportive community of staff, students and visitors, many of whom have provided encouragement and thoughtful feedback during the course of this project. I am especially grateful to Malcolm Allbrook, Nicholas Brown, Maria Haenga-Collins, William Scates Frances, Benjamin Jones, Daniel May, Annemarie McLaren, Maria Nugent and Jayne Regan at the School of History, as well as Diane Erceg and Lilian Pearce at the ANU Fenner School of Environment and Society, for their words of encouragement as I developed my ideas for the book.

A number of other institutions have also provided me with wonderful support and assistance. Cameron Slatyer and research staff at the Australian Museum, in Sydney, shared their knowledge and gave me access to the museum's natural history research collections. Elle Leane, Penny Edmonds, Annalise Rees and other members of the Oceanic Cultures and Connections research group at the University of Tasmania's Institute for Marine and Antarctic Studies shared their scholarly enthusiasm and provided many opportunities for cross-disciplinary conversations about the Southern Ocean. I am especially grateful to them for hosting my visit to Hobart and extending a warm welcome and hospitality during my stay.

Other colleagues and friends in Australia and overseas have contributed to this book in important ways: Alessandro Antonello, Laura Back, Candice and Donovan Edye, Andrea Gaynor, Pete Gill, Heather Goodall, Cathrine Harboe-Ree, Charne Lavery, Cornelia Lüdecke, Mark McGrouther, Narelle McGlusky, Alex Marsden, Steve Nicol, Johanna Parker, Mike Pearson, Steve Rintoul, Helen Rozwadowski and Liz Truswell. Many others have expressed interest in my research on the Southern Ocean over the past few years, and I have appreciated every conversation even though I cannot list them all.

I am grateful to the G Adventures expedition team led by Jonathan Green, as well as the passengers and crew aboard MS *Expedition*, who shared my enthusiasm for the Southern Ocean and its marine inhabitants and contributed to the voyage of a lifetime. Sources of historical information about the Southern Ocean are held by many institutions worldwide, and I am indebted to staff at the Australian National Maritime Museum, National Library of Australia, State Library of New South Wales's Mitchell

Library, Tasmanian Museum and Art Gallery, US Library of Congress, National Oceanic and Atmospheric Administration, Alexander Turnbull Library of New Zealand, National Geographic and Scott Polar Research Institute amongst others for providing invaluable assistance in accessing archival documents, maps and images. A travel scholarship provided by the American Society for Environmental History enabled me to participate in a conference in Toronto and discuss my ideas with an international audience during the formative stages of the project. I would like to acknowledge the Australian Department of the Senate and Parliamentary Library for enabling me to take leave to complete this project, and in particular I thank Cathy Madden and Brien Hallett for their support.

It has been a delight to work with Elspeth Menzies, Paul O'Beirne and the rest of the team at NewSouth Publishing. I am grateful for their professionalism and enthusiastic support for this book, as well as for Penny Mansley's astute copyediting. Of course, I could not have completed this work without the love and encouragement of my family. I am forever grateful to Alan and Lucy for their steadfast support, despite the many interruptions to family life that such a project entails, and to my sister, Jenny, for always being there and believing in my dreams.

NOTES

Prelude

1 The *Ship of Sulaimān* is an account of a Persian embassy that voyaged to Siam (Thailand) in the late seventeenth century. Muḥammad Rabīʿ ibn Muḥammad Ibrāhīm, *The Ship of Sulaimān*, trans. John O'Kane, vol. 49, Routledge, Abingdon, 2007 (1685), p. 163. The manuscript is held in the British Museum, London, BM Or. 6942.

2 As the introductory map shows, the Southern Ocean's geographical limits have long been a matter of disagreement amongst maritime nations. The International Hydrographic Bureau (now Organisation), formed by maritime nations in 1921 to create a uniform system of nautical maps and charts, issued the first edition of *Limits of Oceans and Seas* in 1928 showing the Southern Ocean extending to the southern coastlines of South America, Australia, New Zealand and South America. In the second edition published in 1937, the northern boundary had been moved southward and, by the third edition in 1953, the Southern Ocean was erased altogether and nations advised to determine their own boundary. The United States, for example, adopted the line of latitude at 60° South and Great Britain latitude 55° South, while Australia defined the Southern Ocean as all the waters south of its coastline from latitude 40° South to the Antarctic continent. The Southern Ocean was declared the world's fifth ocean in 2000 in recognition of its physically and biologically distinctive waters south of the Antarctic Convergence, and the IHO issued a draft boundary in 2002, but its northern limits remain contested.

3 Elisabeth Mann Borgese, *The Oceanic Circle: Governing the Seas as a Global Resource*, United Nations University Press, Tokyo, 1998, p. 6.

4 Simon Thomas, 'How the oceans got their names', *Oxford Dictionaries* (blog), 8 June 2015, accessed 10 June 2017, <blog.oxforddictionaries.com/2015/06/water-water-everywhere-ocean-names/>; Peter Heylyn, *Cosmographie in Four Bookes, Containing the Chorographie and Historie of the Whole World, and All the Principall Kingdomes, Provinces, Seas and Isles thereof*, H Seile, London, 1652.

5 Juliet Cummins and David Burchell, *Science, Literature and Rhetoric in Early Modern England*, Ashgate, Farnham, 2007, pp. 191–93; 'Southern Ocean', *Oxford English Dictionary*, accessed 4 December 2016, <www.oed.com/view/Entry/293828?redirectedFrom=Southern+Ocean#eid> (requires subscription).

6 In exploring the idea of the language of the Southern Ocean, I am indebted to the insights on language and the Australian landscape contained in Tim

Bonyhady and Tom Griffiths (eds), *Words for Country: Landscape and Language in Australia*, UNSW Press, Sydney, 2002.

7 Henry M Stommel, *Science of the Seven Seas*, Cornell Maritime Press, New York, 1945, p. ix.

1 Ocean

1 'The seafarer', in Codex Exoniensis*: A Collection of Anglo-Saxon Poetry*, trans. Benjamin Thorpe, Society of Antiquaries, London, 1842 (eleventh century), fols 81v–83r.

2 The five capes are Cape Horn, South America; Cape of Good Hope, South Africa; Cape Leeuwin and South East Cape, Australia; and South West Cape, New Zealand. The Roaring Forties are between latitudes 40° South and 50° South. 'Bluff', in AH McLintock (ed.), *An Encyclopedia of New Zealand*, vol. 1, Government Printer, Wellington, 1966, pp. 213–14.

3 Henry David Thoreau, *Cape Cod*, Ticknor & Fields, Boston, 1865, pp. 99–100, 148.

4 The remaining water is either frozen in glaciers and ice caps or available as fresh water. National Ocean Service, 'How much water is in the ocean?', National Oceanic and Atmospheric Administration, Department of Commerce (US), accessed 7 December 2016, <www.oceanservice.noaa.gov/facts/oceanwater.html>.

5 Simon Thomas, 'How the oceans got their names', *Oxford Dictionaries* (blog), 8 June 2015, accessed 10 June 2017, <blog.oxforddictionaries.com/2015/06/water-water-everywhere-ocean-names/>.

6 'Oceanus', *Encyclopaedia Britannica*, 30 November 2007, accessed 24 August 2016, <www.britannica.com/topic/Oceanus>.

7 Alain Corbin, *The Lure of the Sea: The Discovery of the Seaside in the Western World, 1750–1840*, trans. Jocelyn Phelps, University of California Press, Berkeley, 1994, p. 12.

8 Corbin, *The Lure of the Sea*, p. 12.

9 The ocean has particular spiritual and cultural significance for different cultural groups. See, for example, Ministry for the Environment, Manatū Mō Te Taiao (New Zealand), 'The importance of oceans to New Zealand', reviewed 2 August 2016, accessed 29 December 2017, <mfe.govt.nz/marine/marine-pages-kids/importance-oceans-nz>; National Oceans Office (Australia), *Sea Country: An Indigenous Perspective*, South-East Regional Marine Plan Assessment Report, Hobart, 2002.

10 Thomas Burnet, *The Sacred Theory of the Earth, Containing an Account of the Original of the Earth, and of All the General Changes Which It Hath Already Undergone, or Is to Undergo, Till the Consummation of All Things*, J Hooke, London, 1719 (1681), cited in Corbin, *The Lure of the Sea*, p. 4.

11 Wilfried Jokat, Tobias Boebel, Matthias König and Uwe Meye, 'Timing and geometry of early Gondwana breakup', *Journal of Geophysical Research*, vol. 108, no. B9, September 2003, accessed 2 November 2016, <onlinelibrary.wiley.com/doi/10.1029/2002JB001802/full>; Reg Morrison, *The Voyage of the Great Southern Ark*, Landsdowne, Sydney, 1988, pp. 184–97; Graeme Davison, John

Hirst and Stuart Macintyre (eds), *The Oxford Companion to Australian History*, Oxford University Press, Melbourne, 1998, p. 286.

12 Some sources attribute the renaming of the cape to Bartolomeu Dias himself. Dias drowned in 1500 when his ship was lost at sea near the Cape of Good Hope during a storm. Harold V Livermore, 'Bartolomeu Dias', *Encyclopaedia Britannica*, updated 22 October 2008, accessed 14 July 2017, <www.britannica.com/biography/Bartolomeu-Dias>.

13 William Dampier, *A New Voyage Round the World*, J Knapton, London, 1697, p. 351.

14 George Anson, *A Voyage Round the World in the Years MDCCXL, I, II, III, IV*, ed. Richard Walter and Benjamin Robins, Open University Press, New York, 1974 (1749), p. 84.

15 Ronald S Love, *Maritime Exploration in the Age of Discovery, 1415–1800*, Greenwood, Westport, 2006, p. 88.

16 Corbin, *The Lure of the Sea*, p. 11.

17 Hugh Murray (ed.), *Adventures of British Seamen in the Southern Ocean, Displaying the Striking Contrasts Which the Human Character Exhibits in an Uncivilized State*, Constable, Edinburgh, 1827, p. v.

18 *Declaration Respecting Maritime Law*, Paris, signed and entered into force 16 April 1856, LXI BSP 155.

19 George Shelvocke, *A Voyage Round the World, by Way of the Great South Sea: Performed in a Private Expedition During the War, Which Broke Out with Spain, in the Year 1718*, 2nd edn, M & T Longman, London, 1757.

20 Daniel Defoe, *The Life and Strange Surprizing Adventures of Robinson Crusoe, of York, Mariner, Who Lived Eight and Twenty Years, All Alone in an Un-inhabited Island on the Coast of America, near the Mouth of the Great River of Oroonoque, Having Been Cast on Shore by Shipwreck, Wherein All the Men Perished but Himself, with an Account How He Was at Last as Strangely Deliver'd by Pyrates*, W Taylor, London, 1719.

21 Shelvocke, *A Voyage Round the World*, 2nd edn, pp. 216–18, 265. See also Dampier, *A New Voyage Round the World*; John Bach, 'Dampier, William (1651–1715)', *Australian Dictionary of Biography*, Australian National University, first published 1966, accessed 14 January 2017, <adb.anu.edu.au/biography/dampier-william-1951>.

22 Claudius Ptolemy, *Cosmographia*, trans. Giacomo d'Angelo da Scarperia, Dominicus de Lapis, Bologna, 1477 (c. 150 CE).

23 Michael Pearson, *Great Southern Land: The Maritime Exploration of* Terra Australis, Department of the Environment and Heritage (Australia), Canberra, 2005, p. 3.

24 Mercedes Maroto Camino, *Producing the Pacific: Maps and Narratives of Spanish Exploration (1567–1606)*, Rodopi, Amsterdam, 2005, p. 41.

25 The city of Amsterdam commemorated the sighting with two brass hemispheres inlaid in the white marble floor of the Citizens' Hall, built in 1648–65. In 2013 the Netherlands named its first national scientific laboratory in Antarctica after Dirck Gerritsz. Dick A van der Kroef, 'Gerritsz, Dirck (a.k.a. Dirck Gerritszoon

Pompor) (1544–1608)', in Andrew J Hund, *Antarctica and the Arctic Circle: A Geographic Encyclopedia of the Earth's Polar Regions*, vol. 2, ABC-CLIO, California, 2014, p. 305; Netherlands Organisation for Scientific Research, 'NWO opens first Dutch laboratory on Antarctica', 27 January 2013, accessed 27 January 2018, <www.nwo.nl/en/news-and-events/news/2013/nwo-opens-first-dutch-laboratory-on-antarctica.html>.

26 John Cawte Beaglehole, *The Life of Captain James Cook*, Stanford University Press, Stanford, 1974, p. 113. For a more detailed history of the search for the southern continent, see, for example, William Lawrence Eisler, *The Furthest Shore: Images of* Terra Australis *from the Middle Ages to Captain Cook*, Cambridge University Press, Cambridge, 1995.

27 Bouvet's 'southern land' was in fact a subantarctic island. At latitude 54° South it is regarded as the most remote island in the world. Norway eventually claimed it as a potential whaling station and annexed it by royal decree in 1930. It became a nature reserve in 1971. Alexander Dalrymple, *A Collection of Voyages Chiefly in the Southern Atlantick Ocean*, J Nourse, London, 1775, pp. 3–5.

28 Gerardus Mercator was a Flemish geographer and cartographer who, in 1569, created a cylindrical map projection of the world that converted the globe into a two-dimensional map in such a way that all the parallel lines of latitude had the same length as the equator. The projection enabled the layout of latitude and longitude on Earth's curved surface to be represented on nautical charts, although it distorted the size of landmasses as their distance from the equator increased. It laid the foundation for modern maritime maps. John Cawte Beaglehole, *The Exploration of the Pacific*, A & C Black, London, 1934, p. 9. See also Nicholas Crane, *Mercator: The Man Who Mapped the Planet*, Weidenfeld & Nicolson, London, 2002.

29 John Harris, Navigantium atque itinerantium bibliotheca*; or, A Complete Collection of Voyages and Travels, Consisting of above Six Hundred of the Most Authentic Writers*, rev. edn, ed. John Campbell, 2 vols, T Woodward, etc., London, 1744–8 (first published as Navigantium atque itinerantium bibliotheca*; or, A Compleat Collection of Voyages and Travels, Consisting of above Four Hundred of the Most Authentick Writers*, 2 vols, T Bennett, etc., London, 1705).

30 Beaglehole, *The Life of Captain James Cook*, pp. 118–19, 149.

31 'Secret instructions for Lieutenant James Cook appointed to command His Majesty's bark the *Endeavour*', 30 July 1768, National Library of Australia, Canberra, MS 2.

32 James Cook, *A Voyage Towards the South Pole, and Round the World, Performed in His Majesty's Ships the* Resolution *and* Adventure, *in the Years 1772, 1773, 1774, and 1775*, vol. 1, bk 1, June 1772, W Strahan & T Cadell, London, 1777.

33 Steve SD Jones, William Edward May, Michael William Richey, Tom S Logsdon, John Lawrance Howard and Edward W Anderson, 'Navigation technology', *Encyclopaedia Britannica*, accessed 27 January 2018, <www.britannica.com/technology/navigation-technology>.

34 Lincoln Paine, *The Sea and Civilization: A Maritime History of the World*, Alfred A Knopf, New York, 2013, p. 504.

35 Eóin Phillips, 'Observations on HMS *Resolution*', 1772–5, Papers of the Board of Longitude, Royal Greenwich Observatory Archives, Cambridge, RGO 14/59; Cook, *A Voyage Towards the South Pole*; Dava Sobel, *Longitude: The True Story of a Lone Genius Who Solved the Greatest Scientific Problem of His Time*, Walter, North Mankato, 1995, p. 150.
36 Cook, *A Voyage Towards the South Pole*, bk 1, December 1772.
37 Waves are measured by averaging the height of 15 to 20 waves over a ten-minute period. The buoy was installed by MetOcean Solutions to provide more accurate information about the interaction between waves, atmosphere and ocean in the extreme conditions of the Southern Ocean, in order to improve weather forecasting in the region. Elaine Bunting, 'Southern Ocean storms are some of the most brutal forces in nature: here is how sailors try to harness them', *Yachting World*, 3 February 2015, accessed 25 January 2016, <www.yachtingworld.com/blogs/elaine-bunting/southern-ocean-storms-brutal-forces-nature-sailors-try-harness-61747>; Les Morrow, 'The Southern Ocean factor', *Australian Antarctic Magazine*, no. 4, spring 2002, p. 33; Jamie Morton, 'Giant 19.4 m wave recorded in Southern Ocean', *New Zealand Herald*, 21 May 2017, accessed 2 January 2018, <www.nzherald.co.nz/nz/news/article.cfm?c_id=1&objectid=11860343>.
38 Lionel Arthur Gilbert, 'Banks, Sir Joseph (1743–1820)', *Australian Dictionary of Biography*, Australian National University, first published 1966, accessed 17 January 2017, <adb.anu.edu.au/biography/banks-sir-joseph-1737>; Tom Iredale, 'Forster, Johann Reinhold (1729–1798)', *Australian Dictionary of Biography*, Australian National University, first published 1966, accessed 17 January 2017, <adb.anu.edu.au/biography/forster-johann-reinhold-2057>; Georg Forster, *A Voyage Round the World in His Britannic Majesty's Sloop,* Resolution, *Commanded by Capt. James Cook, During the Years 1772, 3, 4, and 5*, vol. 2, B White, etc., London, 1777. Georg based his account on his father's journals.
39 Michael Dettelbach, '"A kind of Linnaean being": Forster and eighteenth-century natural history', in Johann Reinhold Forster, *Observations Made During a Voyage Round the World*, ed. Nicholas Thomas, Harriet Guest and Michael Dettelbach, University of Hawaii Press, Honolulu, 1996, p. lx (pp. lv–lxxiv) (first published as *Observations Made During a Voyage Round the World, on Physical Geography, Natural History, and Ethic Philosophy, Especially on 1. The Earth and Its Strata, 2. Water and the Ocean, 3. The Atmosphere, 4. The Changes of the Globe, 5. Organic Bodies, and 6. The Human Species*, G Robinson, London, 1778).
40 Cited in Glyn Williams, *Naturalists at Sea: Scientific Travellers from Dampier to Darwin*, Yale University Press, New Haven, 2013, p. 57.
41 G Forster, *A Voyage Round the World*, p. 78.
42 G Forster, *A Voyage Round the World*, p. 420 n.
43 JR Forster, *Observations Made During a Voyage*, p. 414 (original emphasis).
44 JR Forster, *Observations Made During a Voyage*, pp. 69–70, 239.
45 Cook, *A Voyage Towards the South Pole*, bk 2, January 1774.
46 Forster was referring to the argument made by Charles de Brosses in *Histoire des*

navigations aux terres australes, Durand, Paris, 1756. Cited in Dettelbach, 'A kind of Linnaean being', p. lv.

47 The theory of continental drift was first proposed by German meteorologist Alfred Wegener, in 1912, and published as *Die Entstehung der Kontinente und Ozeane* [The origin of continents and oceans], Vieweg, Braunschweig, 1915. Alexander Du Toit's *Our Wandering Continents: An Hypothesis of Continental Drifting*, Oliver & Boyd, Edinburgh, 1937, built on Wegener's theory and became the basis for the subsequent development of the theory of plate tectonics. John Rivett, *An Introduction to Geography, Astronomy, and the Use of the Globes to Which Are Added a Chronological Table of Remarkable Events, Discoveries and Inventions from the Creation to the Year 1794, and a Large Collection of Questions Designed for the Use of Young Persons*, Crouse, Stevenson & Matchett, Norwich, 1794, pp. 56, 58. See also Du Toit, *Our Wandering Continents*, p. 289.

48 Beaglehole, *The Life of Captain James Cook*, p. 107.

49 George ER Deacon, *The Antarctic Circumpolar Ocean*, Cambridge University Press, Cambridge, 1984, p. 1.

50 Helen Rozwadowski, *Fathoming the Ocean: The Discovery and Exploration of the Deep Sea*, Belknap Press of Harvard University Press, Cambridge, 2009, p. 215.

51 Cook, *A Voyage Towards the South Pole*, bk 2, January 1774.

52 Rachel Carson, *The Sea Around Us*, Oxford University Press, Oxford, 1951, p. 194.

2 Wind

1 Tom Griffiths, *Slicing the Silence: Voyaging to Antarctica*, NewSouth, Sydney, 2007, p. 38.

2 Denis Wrigley, *The Wind That Blew Too Much*, Pergamon, Sydney, 1976.

3 The name 'albatross' may have originated from the Spanish or Portuguese *alcatraz*, meaning 'pelican'. Storm tracks are narrow zones along which storms travel, driven by the prevailing winds. Mary E Gillham, *Sub-antarctic Sanctuary: Summertime on Macquarie Island*, AH & AW Reed, Wellington, 1967, p. 20; Mary E Gillham, *Island Hopping in Tasmania's Roaring Forties*, Arthur H Stockwell, Ilfracombe, 2000; Mary Gillham Archive Project, 'Mary Gillham biography', accessed 10 July 2017, <marygillhamarchiveproject.com/biography/>; Bernadette Hince, *The Antarctic Dictionary: A Complete Guide to Antarctic English*, CSIRO, Canberra, 2000, pp. 2–3; William Lancelot Noyes Tickell, *Albatrosses*, Pica, Mountfield, East Sussex, 2000, p. 14; Jaimie Cleeland, 'The women of Macquarie Island', Australian Antarctic Division, Department of the Environment and Energy (Australia), 28 March 2014, accessed 10 July 2017, <antarctica.gov.au/living-and-working/stations/macquarie-island/this-week-at-macquarie-island/2014/28-march-2014/3>.

4 The species name, *exulans*, of the wandering albatross (*Diomedea exulans*) means 'exiled' or 'wandering'. The *IOC World Bird List* notes that the species names for albatrosses are controversial. Frank Gill and David Donsker (eds), 'Loons, penguins, petrels', *IOC World Bird List*, version 8.1, 25 January 2018, accessed 28 January 2018, <www.worldbirdnames.org/bow/loons/>.

5 Graham Barwell, 'What's in a name? What names for albatross genera reveal

about attitudes to the birds', *Animal Studies Journal*, vol. 7, no. 1, 2012, pp. 68–70.

6 Tom Garrison, *Oceanography: An Invitation to Marine Science*, Brooks / Cole-Thomson Learning, Belmont, 2005, p. 365.

7 Charles Waitara, funeral oration for Te Whiti, 1907, cited in Ana Pallesen, '*Roimata toroa* (tears of the albatross): a historical review of the albatross in folklore, and a critical examination of the environmental law protections', Graduate Certificate literature review, University of Canterbury, Christchurch, 2008. See also Paul Sagar, 'Albatrosses', Te Ara*: The Encyclopedia of New Zealand*, reviewed and revised 17 February 2015 (12 June 2006), accessed 11 July 2017, <www.teara.govt.nz/en/albatrosses/page-1>.

8 Robert Cushman Murphy, *Logbook for Grace: Whaling Brig Daily, 1912–1913*, R Hale, London, 1948, p. 116. Murphy is best known for his book *Oceanic Birds of South America: A Study of Species of the Related Coasts and Seas, Including the American Quadrant of Antarctica*, Macmillan, New York, 1936.

9 Gillham, *Sub-antarctic Sanctuary*, p. 37; Samuel Taylor Coleridge, *The Rime of the Ancient Mariner*, in William Wordsworth and Samuel Taylor Coleridge, *Lyrical Ballads, with a Few Other Poems*, vol. 1, J & A Arch, London, 1798, pp. 15–63.

10 Seamus Perry, 'An introduction to *The Rime of the Ancient Mariner*', British Library, 15 May 2014, accessed 28 January 2018, <www.bl.uk/romantics-and-victorians/articles/an-introduction-to-the-rime-of-the-ancient-mariner>.

11 The Australian historian Bernard Smith suggested that Samuel Taylor Coleridge may have been influenced in part by stories told by William Wales, who served as James Cook's mathematician and astronomer on board the *Resolution*. George Shelvocke, *A Voyage Round the World by Way of the Great South Sea, Perform'd in the Years 1719, 20, 21, 22, in the* Speedwell *of London*, J Senex, London, 1726, pp. 72–73; Smith, *Imagining the Pacific: In the Wake of the Cook Voyages*, Yale University Press, New Haven, 1992, p. x.

12 Conservation measures have been introduced aimed at reducing seabird mortality from longline fishing, including setting longlines at night, reducing offal discharge from fishing vessels and promoting the use of bird-saving devices. International Union for Conservation of Nature, '*Diomedea exulans*', *IUCN Red List of Threatened Species*, 2017, accessed 13 July 2017, <www.iucnredlist.org/details/classify/22698305/0>.

13 The *Tryall*'s crew became the first British sailors to record a sighting of Australia. Their vessel is Australia's oldest known shipwreck.

14 In 1790 the *Sirius* was wrecked on a reef while landing stores on Norfolk Island. The crew were rescued and returned to Britain, while Hunter went on to serve as the governor of New South Wales between 1795 and 1799. The wreck of the *Sirius* is now protected under Australia's *Historic Shipwrecks Act 1976*. John Hunter, *An Historical Journal of the Transactions at Port Jackson and Norfolk Island with the Discoveries Which Have Been Made in New South Wales and in the Southern Ocean, since the Publication of Phillip's Voyage, Compiled from the Official Papers, Including the Journals of Governor Phillip and King, and of Lieut. Ball, and*

the Voyages of the First Sailing of the Sirius *in 1787, to the Return of That Ship's Company to England in 1792*, J Stockdale, London, 1793.

15 Henry David Thoreau, *Cape Cod*, Ticknor & Fields, Boston, 1865, pp. 98–99.

16 Mark Twain, *Life on the Mississippi*, JR Osgood, Boston, 1883, p. 119.

17 Jonathan Raban, *Passage to Juneau: A Sea and Its Meanings*, Picador, Surrey, 1999, p. 94; Greg Dening, 'Deep times, deep spaces: civilizing the sea', in Bernhard Klein and Gesa Mackenthun (eds), *Sea Changes: Historicizing the Ocean*, Routledge, Abingdon, 2004, pp. 15–16, 33 (pp. 13–36).

18 John MacDonald, *Travels, in Various Parts of Europe, Asia, and Africa, During a Series of Thirty Years and Upwards*, Forbes, London, 1790, p. 276; Coleridge, *The Rime of the Ancient Mariner*; Walter Scott, *Rokeby*, J Ballantyne, Edinburgh, 1813; Richard Wagner, *Der fliegende Holländer* [The flying Dutchman], opera in three acts, Germany, 1843; Albert Pinkham Ryder, *Flying Dutchman*, by 1887, oil on canvas, 36.1 × 43.8 cm, Smithsonian American Art Museum, 1929.6.95.

19 The work was ascribed to George Barrington, who was described as 'Superintendent of the Convicts at Parramatta', although the *Encyclopaedia Britannica* notes that there is no evidence that he wrote it. Indiamen, or, more accurately, East Indiamen, were sailing ships operating under charter to Danish, Dutch, English, French, Portuguese or Swedish companies trading between Europe and India, with some extending their routes to China. Barrington, *Voyage to Botany Bay, with a Description of the Country, Manners, Customs, Religion, &c. of the Natives*, HD Symonds, London, 1795, pp. 45–46; 'George Barrington', *Encyclopaedia Britannica*, updated 2 February 2007, accessed 4 February 2018, <www.britannica.com/biography/George-Barrington>.

20 John N Dalton (ed.), *The Cruise of H.M.S. 'Bacchante', 1879–1882*, Macmillan, London, 1886, p. 551.

21 The Australian Hydrographic Service defines the strait as part of the Tasman Sea rather than of the Southern Ocean. See Australian Hydrographic Service, *Names and Limits of Oceans and Seas Around Australia*, AHS AA609582, 2012, Australian Hydrographic Office, Department of Defence (Australia), December 2017, accessed 20 February 2018, <hydro.gov.au/factsheets/WFS_Names_and_Limits_of_Oceans_and_Seas_Around_Australia.pdf>.

22 Patsy Cameron, 'Aboriginal life pre-invasion', in Alison Alexander (ed.), *The Companion to Tasmanian History*, Centre for Tasmanian Historical Studies, University of Tasmania, 2006, accessed 2 February 2018, <www.utas.edu.au/library/companion_to_tasmanian_history/A/Aboriginal life pre-invasion.htm>; Iain Davidson and David Roberts, 'On being alone: the isolation of Tasmania', in Martin Crotty and David Roberts (eds), *Turning Points in Australian History*, UNSW Press, Sydney, 2009, pp. 18–31.

23 Department of the Environment and Energy (Australia), 'Australian national shipwrecks database', 2009, accessed 18 July 2017, <www.environment.gov.au/heritage/historic-shipwrecks/australian-national-shipwreck-database>.

24 The Cape Otway Lighthouse was built in 1846 after hundreds of lives were lost in such shipwrecks along the coastlines of Bass Strait. In 2006 researchers calculated that there had been 836 severe storms in the strait – an average of

6 per year – since measurements began to be recorded from the lighthouse, in 1865. 'Dreadful shipwreck: wreck of the "Cataraqui" emigrant ship of 800 tons; 414 lives lost', *Courier* (Hobart), 20 September 1845, p. 2; Lisa V Alexander and Scott Power, 'Severe storms inferred from 150 years of sub-daily pressure observations along Victoria's "shipwreck coast"', *Australian Meteorological and Oceanographic Journal*, vol. 48, 2009, pp. 129–33.

25 Henrik Johan Bull, *The Cruise of the 'Antarctic' to the South Polar Regions*, E Arnold, London, 1896, pp. 44–45.

26 This series of hydrographic charts for the Atlantic, Pacific and Indian oceans was produced under the auspices of the US Hydrographic Office and is now held in the American Geographical Society Library collection. Matthew Fontaine Maury, 'Maury's wind and currents chart 1852', 1852, American Geographical Society Library, University of Wisconsin-Milwaukee Libraries, Milwaukee, (AGS) (RARE) 700 A-Var Maury Charts.

27 Matthew Fontaine Maury, *The Physical Geography of the Sea*, 3rd edn, Harper, New York, 1855, pp. vii, 292; Eric P Chassingnet and Jacques Vernon (eds), *Ocean Weather Forecasting: An Integrated View of Oceanography*, Springer, Dordrecht, 2006, p. 3.

28 Joseph-Antoine Raymond de Bruni D'Entrecasteaux led an expedition to search for Jean François de Galaup de La Pérouse, whose vessel vanished after leaving Botany Bay in 1788. He sailed from Cape Town to the Pacific Ocean via the southern coast of Australia, completing a detailed survey of the coast of Tasmania along the way. Louis-Claude Desaulses de Freycinet was a cartographer-surveyor with the French expedition led by Nicolas Thomas Baudin to survey the Australian coastline in 1802. He subsequently commanded his own survey vessel to chart parts of the southern coastline.

29 Most nautical charts used the old Mercator navigation method developed, in 1569, by Gerardus Mercator. Samuel Birley Rowbotham, 'Great circle sailing', in *Zetetic Astronomy: Earth Not a Globe; An Experimental Inquiry into the True Figure of the Earth, Proving It a Plane, without Orbital or Axial Motion, and the Only Known Material World, Its True Position in the Universe, Comparatively Recent Formation, Present Chemical Condition, and Approaching Destruction by Fire, &c., &c., &c. by 'Parallax'*, 3rd edn, Simpkin, Marshall, London, 1881, p. 283.

30 John Thomas Towson, *The Principles of Great Circle and Composite Sailing: A Paper Read Before the Literary and Philosophical Society, 1853, with Observations Made from More Recent Voyages*, H Greenwood, Liverpool, 1855, p. 11.

31 The origin of the term 'clipper' is unclear, but it may have been adapted from the English word 'clip', meaning 'to run or fly swiftly'. Arthur H Clarke, *The Clipper Ship Era: An Epitome of Famous American and British Clipper Ships, Their Owners, Builders, Commanders, and Crews, 1843–69*, GP Putnam's Sons, New York, 1912; 'Clipper ship', *Encyclopaedia Britannica*, 20 July 1998, accessed 15 July 2017, <www.britannica.com/technology/clipper-ship>.

32 The barque *George Marshall* was wrecked off Flinders Island, in the notorious Bass Strait passage, in 1862. Australian National Maritime Museum, 'Passenger ships to Australia: a comparison of vessels and journey times to Australia

between 1788 and 1900', accessed 30 December 2017, <www.anmm.gov.au/Learn/Library-and-Research/Research-Guides/Passenger-Ships-to-Australia-A-Comparison-of-Vessels-and-Journey-Time>.

33 The *City of Adelaide*'s final return voyage took place in 1887. The vessel is the world's oldest surviving composite clipper ship (built with a wooden hull on iron frames). Local newspapers routinely reported the arrivals and departures of ships in Australian ports. See the National Library of Australia's newspaper collection online in the *Trove* database, 'Digitised newspapers and more', *Trove*, National Library of Australia, accessed 17 January 2018, <trove.nla.gov.au/newspaper/?q=>.

34 Towson, *The Principles of Great Circle and Composite Sailing*, pp. 16–17.

35 John Thomas Towson, *Icebergs in the Southern Ocean: A Paper Read Before the Historic Society of Lancashire and Cheshire, on the 19th November 1857, with Observations Made from More Recent Reports*, T Brakell, Liverpool, 1859.

36 'Sailing directions for the south coast of Australia', in Great Britain, Hydrographic Department, *The Australia Directory*, vol. 1; *Containing Directions for the Southern Shores of Australia from Cape Leeuwin to Port Stephens, Including Bass Strait and Van Diemen's Land*, Hydrographical Office, Admiralty, London, 1830, p. 1 (pp. 1–93).

37 Square riggers were sailing ships with large square sails carried on horizontal beams that were perpendicular to the mast and keel of the vessel. The *Great Eastern*'s passenger service was short lived, and it was eventually used to lay undersea telegraph cables and, finally, as a floating amusement park. Royal Museums Greenwich, 'Ships and steam power', accessed 29 August 2017, <www.rmg.co.uk/discover/explore/steam-power>.

38 Joshua Slocum, *The Voyages of Joshua Slocum*, ed. Walter Magnes Teller, Rutgers University Press, New Brunswick, 1958, pp. 209, 278; Charles Darwin, *Journal and Remarks, 1832–1836*, vol. 3 of Robert FitzRoy (ed.), *Narrative of the Surveying Voyages of His Majesty's Ships* Adventure *and* Beagle, *Between the Years 1826 and 1836, Describing Their Examination of the Southern Shores of South America, and the* Beagle*'s Circumnavigation of the Globe*, H Colburn, London, 1839, p. 307. Darwin's journal, now known as *The Voyage of the* Beagle, was published in 1839 as *Journal and Remarks, 1832–1836*. A revised, 2nd edition was published in 1845, as *Journal of Researches into the Natural History and Geology of the Various Countries Visited During the Voyage of H.M.S.* Beagle *Round the World under the Command of Capt. FitzRoy, R.N.*, J Murray, London, 1845, and was followed by several subsequent editions. The version cited here is an online scanned copy of *Journal and Remarks*: John van Whye (ed.), *The Complete Works of Charles Darwin Online*, updated 2 July 2012, accessed 15 February 2018, <darwin-online.org.uk/content/frameset?pageseq=1&itemID=F10.3&viewtype=text>.

39 Charles Doane, 'Bernard Moitessier: sailing mysticism and *The Long Way*', *Wavetrain*, 3 September 2015, accessed 30 December 2017, <wavetrain.net/lit-bits/689-bernard-moitessier-sailing-mysticism-and-the-long-way>.

40 Bernard Moitessier, *The Long Way*, Sheridan House, New York, 1995, p. 141.

41 Kay Cottee, *First Lady: A History-making Solo Voyage Around the World*, Macmillan, South Melbourne, 1989, pp. 69, 77, 131, 136, 142.

3 Coast

1 Derek Mahon, 'The banished gods', 1975, in *Collected Poems*, Gallery, Oldcastle, 1999, p. 85, by kind permission of the author and The Gallery Press, <www.gallerypress.com>.

2 Thierry Aubin, Pierre Jouventin and Christophe Hildebrand, 'Penguins use the two-voice system to recognize each other', *Proceedings of the Royal Society B*, vol. 267, no. 1448, 7 June 2000, pp. 1081–87.

3 Tom Griffiths, *Slicing the Silence: Voyaging to Antarctica*, NewSouth, Sydney, 2007, pp. 174, 4.

4 Robert FitzRoy (ed.), *Narrative of the Surveying Voyages of His Majesty's Ships* Adventure *and* Beagle, *Between the Years 1826 and 1836, Describing Their Examination of the Southern Shores of South America, and the* Beagle*'s Circumnavigation of the Globe*, 3 vols, H Colburn, London, 1839.

5 Cited in Claude C Albritton Jr, *The Abyss of Time: Unravelling the Mystery of the Earth's Age*, Freeman, Cooper, Devon, 1980, p. 164.

6 Charles Darwin, *Journal and Remarks, 1832–1836*, vol. 3 of Robert FitzRoy (ed.), *Narrative of the Surveying Voyages of His Majesty's Ships* Adventure *and* Beagle, *Between the Years 1826 and 1836, Describing Their Examination of the Southern Shores of South America, and the* Beagle*'s Circumnavigation of the Globe*, H Colburn, London, 1839, pp. 301–304, 305.

7 Nathan Curtis, 'All aboard the emergent ark: biogeography of the dune insect fauna of New Zealand and Chatham Island', PhD thesis, Lincoln University, Canterbury, 2001, pp. 20–21.

8 Stephen DA Smith, 'Kelp rafts in the Southern Ocean', *Global Ecology & Biogeography*, vol. 11, 2002, pp. 67–69.

9 Bernadette Hince, *The Antarctic Dictionary: A Complete Guide to Antarctic English*, CSIRO, Canberra, 2000, p. 18.

10 Ceridwen I Fraser, Raisa Nikula and Jonathan M Waters, 'Oceanic rafting by a coastal community', *Proceedings of the Royal Society B*, 15 September 2010, accessed 9 October 2017, <rspb.royalsocietypublishing.org/content/roymprsb/early/2010/09/13/rspb.2010.1117.full.pdf>.

11 Anne Chapman, *European Encounters with the Yámana People of Cape Horn, Before and After Darwin*, Cambridge University Press, Cambridge, 2010, p. 167.

12 Cited in Chapman, *European Encounters*, pp. 14, 171.

13 The Yaghan are also known as Yahgan, although, according to the British missionary Thomas Bridges, the Yaghan described themselves as Yamana, meaning 'humanity', in order to distinguish themselves from other indigenous language groups. The Alakaluf are also known as the Kawésqar, Kaweskar, Alacaluf, Alacalufe or Halakwulup, meaning 'mussel eater' in Yaghan; the Selk'nam are also known as Ona; the Haush called themselves Manek'enk. Anne Chapman, *Darwin in Tierra del Fuego*, Imago Mundi, Buenos Aires, 2006, pp. 7–8; Fernanda Peñaloza, Claudio Canaparo and Jason Wilson, *Patagonia:*

Myths and Realities, P Lang, Bern, 2010, p. 113; Martin Gusinde, *The Yámana: The Life and Thought of the Water Nomads of Cape Horn*, Human Relations Area Files, New Haven, 1961 (1937), cited in Chapman, *European Encounters*, p. 695.

14 Arnoldo Canclini, *The Fuegian Indians: Their Life, Habits and History*, Continente, Buenos Aires, 2012, p. 14.

15 During the course of her fieldwork in Tierra del Fuego, Anne Chapman published many books on the indigenous peoples and included rare photographs of Yaghans taken by members of the French expedition to Cape Horn in 1882–3. Chapman, *European Encounters*, p. 172.

16 Cited in Chapman, *European Encounters*, p. 28.

17 Gertrude Himmelfarb, *Darwin and the Darwinian Revolution*, Chatto & Windus, London, 1959, p. 55.

18 Darwin, *Journal and Remarks*, p. 228; Nic Compton, *Off the Deep End: A History of Madness at Sea*, Bloomsbury, London, 2017.

19 Chapman, *European Encounters*, p. 15.

20 A cordillera is a mountain range comprising several parallel chains. The name is derived from the Spanish word *cordilla*, meaning 'cord' or 'little rope'. Darwin, *Journal and Remarks*, pp. 236–37.

21 William Johnson Sollas, *Ancient Hunters and Their Modern Representatives*, Macmillan, London, 1911, p. 393.

22 Charles Wellington Furlong, 'Some effects of environment on the Fuegian tribes', *Geographical Review*, vol. 3, no. 1, January 1917, pp. 3, 10–11 (pp. 1–15); Charles Wellington Furlong, 'Tribal distribution and settlements of the Fuegians', *Geographical Review*, vol. 3, no. 3, March 1917, pp. 169–87.

23 Charles Wellington Furlong, 'The vanishing people of the Land of Fire', *Harper's Magazine*, January 1910, pp. 217–29.

24 Francisco Coloane's books include *Capo de Hornos*, Orbe, Switzerland, 1941; *La Tierra del Fuego*, trans. David Petreman and Howard Curtis, Editorial del Pacifico SA, Santiago, 1956; *Los conquistadores de la Antártida* [The conquerors of Antarctica], Zig-Zag, Santiago, 1945; and *El camino de la ballena* [The whale's path], Zig-Zag, Santiago, 1962.

25 Anne Chapman, 'A genealogy of my professors and informants', *Anne Mackaye Chapman*, 2005, accessed 17 August 2017, <www.thereedfoundation.org/rism/chapman/genealogy.htm>. Chapman's work includes *Drama and Power in a Hunting Society: The Selk'nam of Tierra del Fuego*, Cambridge University Press, Cambridge, 1982; *Darwin in Tierra del Fuego*; *European Encounters*; and Chapman and Ana Montes de Gonzales, *The Ona People: Life and Death in Tierra del Fuego*, documentary film, Documentary Educational Resources, Watertown, MA, 1977.

26 Will Meadows, 'What happened to the fires of Tierra del Fuego?', *National Geographic* (blog), 10 February 2014, accessed 11 February 2018, <blog.nationalgeographic.org/2014/02/10/what-happened-to-the-fires-of-tierra-del-fuego/>.

27 Will Meadows, 'Living embers: Yaghan art, culture, and bone harpoons',

National Geographic (blog), 18 February 2014, accessed 18 August 2017, <blog.nationalgeographic.org/2014/02/18/living-embers-yaghan-art-culture-and-bone-harpoons/>. Permission kindly granted by Will Meadows and the National Geographic Society.

28 Bernadette Hince, 'The teeth of the wind: an environmental history of subantarctic islands', unpublished PhD thesis, Australian National University, Canberra, 2005, p. 9. Courtesy of the author.

29 Geoscience Australia, 'Heard and McDonald Islands', accessed 23 August 2017, <www.ga.gov.au/scientific-topics/national-location-information/dimensions/remote-offshore-territories/heard-and-mcdonald-islands>.

30 Patrick G Quilty noted that while the exposed portions are 'ephemeral', large-scale features that exist under the sea are much older. Quilty, 'Origin and evolution of the sub-antarctic islands: the foundation', *Papers and Proceedings of Royal Society of Tasmania*, vol. 141, no. 1, 2007, p. 54 (pp. 35–58).

31 George A Knox, *Biology of the Southern Ocean*, 2nd edn, CRC, Boca Raton, 2007, p. 210.

32 Hince, 'The teeth of the wind', p. 22.

33 Erica Nathan, *Heard Island Is a Place*, shortlisted essay in Nature Writing Prize 2017, 2017, p. 1, Nature Conservancy Australia, accessed 11 October 2017, <www.natureaustralia.org.au/wp-content/uploads/2017/05/Heard-Island-is-a-Place-by-Erica-Nathan.pdf>.

34 John Cawte Beaglehole, *The Life of Captain James Cook*, Stanford University Press, Stanford, 1974, p. 514.

35 Michael Pearson, Ruben Stehberg, Andrés Zarankin, M Ximena Senatore and Carolina Gatica, 'Conserving the oldest historic sites in the Antarctic: the challenges in managing the sealing sites in the South Shetland Islands', *Polar Record*, vol. 46, no. 1, January 2010, pp. 57–64.

36 Robert Headland served as an officer with the British Antarctic Survey and spent several seasons on South Georgia during the 1970s and 1980s undertaking scientific research and acquiring a wealth of knowledge about the island's natural and cultural history. Headland, *The Island of South Georgia*, Cambridge University Press, Cambridge, 1992, p. 53.

37 Michael Pearson, 'Charting the sealing islands of the Southern Ocean', *Globe*, vol. 80, 2016, pp. 33–56; Pearson et al., 'Conserving the oldest historic sites', p. 58.

38 Fabian Gottlieb Thaddeus von Bellingshausen's expedition was the first to sight the continent of Antarctica. Bellingshausen Basin is named in his honour. Bellingshausen, *The Voyage of Captain Bellingshausen to the Antarctic Seas, 1819–21*, ed. Frank Debenham, trans. from Russian, vol. 2, Routledge, Abingdon, 2017, p. 425; Philip Hoare, *The Sea Inside*, Melville House, New York, 2014, p. 292.

39 'New scientific voyage of discovery', *Australian*, 6 November 1829, p. 4.

40 William Henry Bayley Webster, *Narrative of a Voyage to the Southern Atlantic Ocean, in the Years 1828, 29, 30, Performed in H.M. Sloop* Chanticleer, *under the Command of the Late Captain Henry Foster, F.R.S. &c.*, R Bentley, London, 1834, p. 157.

41 See, for example, Pearson, 'Charting the sealing islands'; Pearson et al., 'Conserving the oldest historic sites'; Rebe Taylor, *Unearthed: The Aboriginal Tasmanians of Kangaroo Island*, Wakefield, Kent Town, 2002; Headland, *The Island of South Georgia*, pp. 51–57.

42 The population had recovered by 1993, when it was estimated to be around 1.5 million. Knox, *Biology of the Southern Ocean*, p. 216.

43 A try-pot was a large cast-iron pot with a rounded bottom and two flattened sides that was heated from below to melt down (try out) blubber which, when heated, would be rendered into oil. Such pots were also used to extract oil from whales and penguins. Australian Antarctic Division, 'Elephant seals', Department of the Environment and Energy (Australia), last updated 9 January 2018, accessed 8 February 2018, <antarctica.gov.au/about-antarctica/wildlife/animals/seals-and-sea-lions/elephant-seals>.

44 See Max Downes, *Indexing Sealers' Logbooks from Heard Island*, ANARE Research Notes no. 97, Australian Antarctic Division, Department of the Environment and Energy (Australia), Kingston, 1996, p. 5.

45 The fossilised woods brought back by the expedition were not examined until 1921, when the British Museum identified them as coniferous. 'Fossil flora of Kerguelen Island', *Nature*, vol. 162, no. 4112, 21 August 1948, p. 309; Henry Nottidge Moseley, *Notes by a Naturalist on the 'Challenger', Being an Account of Various Observations Made During the Voyage of H.M.S. 'Challenger' Round the World, in the Years 1872–1876, under the Commands of Capt. Sir G. S. Nares, R.N., K.C.B., F.R.S., and Capt. F. T. Thomson, R.N.*, Macmillan, London, 1879, pp. 169–70.

46 Joseph Dalton Hooker to Charles Darwin, 29 January 1844, Cambridge University Library, Cambridge, MS DAR 100: 5–7; Joseph Dalton Hooker, *The Botany of the Antarctic Voyage of H.M. Discovery Ships* Erebus *and* Terror *in the Years 1839–1843, under the Command of Captain Sir James Clark Ross, Kt., R.N., F.R.S. &c.*, 6 vols, Reeve, London, 1844–59; Charles Darwin to Joseph Dalton Hooker, [5 or 12 November 1845], Cambridge University Library, Cambridge, MS DAR 114: 45, 45b.

47 Cited in Stephen D Hopper and Hans Lambers, 'Darwin as a plant scientist: a Southern Hemisphere perspective', *Trends in Plant Science*, vol. 14, no. 8, August 2009, pp. 421–35; Charles Darwin to Charles Lyell, 11 February 1857, *Darwin Correspondence Project*, letter no. 2050, University of Cambridge, accessed 10 December 2017, <www.darwinproject.ac.uk/letter/?docId=letters/DCP-LETT-2050.xml;query=11%20February%201857;brand=default;hit.rank=1%20-%20hit.ran>.

48 Patrick G Quilty and Graeme E Wheller, 'Heard Island and the McDonald Islands: a window into the Kerguelen Plateau', *Papers and Proceedings of the Royal Society of Tasmania*, vol. 133, no. 2, 2000, pp. 1–12.

49 Hince, 'The teeth of the wind', p. 23.

50 Cited in Moseley, *Notes by a Naturalist on the 'Challenger'*, p. 225.

51 Cited in Richard Corfield, *The Silent Landscape: The Scientific Voyage of HMS* Challenger, Joseph Henry, Washington, DC, 2003, p. 148.

52 James Cook, *The Three Voyages of Captain James Cook Round the World*, vol. 5, Longman, London, 1821, pp. 165–66; Joseph Matkin, *At Sea with the Scientifics: The* Challenger *Letters of Joseph Matkin*, ed. Philip F Rehbock, University of Hawaii Press, Honolulu, 1992, p. 122; George Campbell, *Log-letters from 'The Challenger'*, Macmillan, London, 1876, p. 81.

53 Bjørn L Basberg and Robert K Headland, 'The economic significance of the 19th century Antarctic sealing industry', *Polar Record*, vol. 49, no. 4, October 2013, pp. 381–91. See also Michael Nash, *The Bay Whalers: Tasmania's Shore-based Whaling Industry*, Navarine, Woden, 2003.

54 Moseley, *Notes by a Naturalist on the 'Challenger'*, p. 179.

55 Matkin, *At Sea with the Scientifics*, p. 133.

56 Circumnavigation Committee of the Royal Society, 'Plans for the expedition, 1872', cited in Charles Wyville Thomson, *The Voyage of the 'Challenger': The Atlantic; A Preliminary Account of the General Results of the Exploring Voyage of H.M.S. 'Challenger' During the Year 1873 and the Early Part of the Year 1876*, Macmillan, London, 1879, p. 71.

57 According to geologists, Macquarie Island probably formed between 30 and 11 million years ago 6 kilometres below Earth's oceanic crust. As the spreading stopped the crust began to compress, squeezing rocks upward to form a ridge. About 600 000 years ago the ridge emerged as an island above the surface of the Southern Ocean. The *World Heritage List* recognised the geological significance of Macquarie Island in 1997. Parks & Wildlife Service (Tasmania), 'Geoheritage', last modified 6 November 2008, accessed 9 September 2017, <www.parks.tas.gov.au/index.aspx?base=620>. See also World Heritage Centre, 'Macquarie Island', *World Heritage List*, United Nations Educational, Scientific and Cultural Organization, accessed 11 October 2017, <whc.unesco.org/en/list/629>.

58 Australian Antarctic Division, 'Macquarie Island station: a brief history', last updated 17 June 2002, accessed 12 February 2018, <antarctica.gov.au/about-antarctica/history/stations/macquarie-island>.

59 Cited in 'Sydney', *Sydney Gazette and New South Wales Advertiser*, 13 December 1822, p. 2.

60 Edward Adrian Wilson was serving as one of the expedition's two doctors, and he accompanied Robert Falcon Scott and Ernest Shackleton on their attempt to reach the South Pole in 1902. He was subsequently one of the five men who perished during Scott's second attempt to reach the pole, in 1911. He was 39 years old. Wilson, *Diary of the* Discovery *Expedition to the Antarctic Regions, 1901–1904*, ed. Ann Savours, Blandford, London, 1966, pp. 77, 78.

61 The Australasian Antarctic Expedition also established bases at Commonwealth Bay and the Shackleton Ice Shelf, on the Antarctic Peninsula, and claimed the land near the bases as British territory. Douglas Mawson, *Macquarie Island: Its Geography and Geology*, Australasian Antarctic Expedition 1911–14, Scientific Reports, series A, vol. 5, Government Printer, Sydney, 1943, p. 15.

62 John Stanley Cumpston, *Macquarie Island*, Australian Antarctic Division, Department of External Affairs (Australia), Canberra, 1968.

63 'Mr Hatch's lecture', *Evening Star* (Dunedin), no. 9807, 21 September 1895, p. 4.

64 See, for example, 'Macquarie Islands', *Tasmanian News* (Hobart), 13 June 1895, p. 2.

65 Cited in Douglas Mawson, *Geographical Narrative and Cartography*, Australian Antarctic Expedition 1911–14, Scientific Reports, series A, vol. 1, pt 1, Government Printer, Sydney, 1942, p. 281.

66 Hatch's lease on the island expired in 1920, and he stood as a Nationalist candidate for the Tasmanian seat of Denison but was not elected. 'Sinking a small fortune: Joseph Hatch and the oiling industry', *The Sealers' Shanty: A Journal of Sealers' Stories*, vol. 9, 1889–1919 (13 April 1917), pp. 1–4.

67 See, for example, Griffiths, *Slicing the Silence*, pp. 128–29.

68 Isobel Bennett began her career by accident, meeting zoologist William Dakin on a cruise and becoming the research assistant for his book *Whalemen Adventures*. She subsequently collaborated with him on his *Australian Seashores*, which she completed after his death, in 1950. Her papers are held in the National Library of Australia manuscripts collection. Bennett, *Shores of Macquarie Island*, Rigby, Adelaide, 1971, pp. 37, 33. See also 'Dr Isobel Bennett (1909–2008), marine biologist', interview by Nessy Allen, Australian Academy of Science, 2000, accessed 11 October 2017, <www.science.org.au/learning/general-audience/history/interviews-australian-scientists/dr-isobel-bennett-1909-2008>.

69 'Dr Isobel Bennett'.

70 Bennett, *Shores of Macquarie Island*, pp. 22, 56–59, 18, 19.

71 In 1997 Macquarie, Heard and McDonald islands were recognised by the *World Heritage List* for their unique geological features and 'rare pristine island ecosystems'. The Australian Government subsequently declared the waters around each island a marine reserve. See World Heritage Centre, 'Heard and McDonald Islands', *World Heritage List*, United Nations Educational, Scientific and Cultural Organization, accessed 12 October 2017, <whc.unesco.org/en/list/577>; World Heritage Centre, 'Macquarie Island'.

72 Bennett, *Shores of Macquarie Island*, pp. 22, 68–69.

73 The Heard Island station was established in December 1947 and the Macquarie Island station in March 1948.

74 Eric Bird (ed.), *Encyclopedia of the World's Coastal Landforms*, Springer, London, 2010, p. 1466.

75 Australian Antarctic Division, 'Human activities', Department of the Environment and Energy (Australia), last modified 28 February 2005, accessed 12 October 2017, <heardisland.antarctica.gov.au/about/human-activities>.

76 Britain's argument for territorial control of Heard Island rested on the claim that a British sealer had been the first to sight the island, in 1833. Australia's claim in 1947 served to confirm its tenure under the British Crown. See Hince, 'The teeth of the wind', pp. 117–19.

77 Arthur Scholes, *Fourteen Men: The Story of the Australian Antarctic Expedition to Heard Island*, G Allen & Unwin, London, 1951, p. 30.

78 Scholes, *Fourteen Men*, p. 50. See also Grahame M Budd, 'Australian exploration of Heard Island, 1947–1971', *Polar Record*, vol. 43, no. 2, April 2007, pp. 97–123.

4 Ice

1 Louis Bernacchi, *To the South Polar Regions: Expedition of 1898–1900*, Bluntisham, Denton, 1991 (1901), p. 31.

2 Herbert Ponting, *The Great White South; or, With Scott in the Antarctic, Being an Account of Experiences with Captain Scott's South Pole Expedition and of the Natural Life of the Antarctic*, Duckworth, London, 1935 (1921), p. 15.

3 Stephen J Pyne, *The Ice: A Journey to Antarctica*, University of Washington Press, Seattle, 1986, p. 2.

4 Richard M Laws, *Large Animals and Wide Horizons: Adventures of a Biologist; The Autobiography of Richard M. Laws*, pt 3, *Antarctica and Academe*, ed. Arnoldus Schytte Blix, Scott Polar Research Institute, Cambridge, 2012, p. 266.

5 Pyne, *The Ice*, p. 74; Mark Prigg and Ellie Zolfagharifard, 'The underworld of the Antarctic: stunning new images reveal what the UNDERNEATH of an iceberg looks like', *Daily Mail Australia*, 17 February 2015, accessed 30 December 2017, <www.dailymail.co.uk/sciencetech/article-2956319/The-underworld-Antarctic-Stunning-new-images-reveal-UNDERNEATH-iceberg-looks-like.html>.

6 '*Nga-iwi-o-aotea*', Te Ao Hou*: The Māori Magazine*, no. 59, June–August 1967, p. 43.

7 Warren Judd, 'Polynesians in the Southern Ocean', *New Zealand Geographic*, no. 69, September–October 2004, accessed 30 December 2017, <www.nzgeo.com/stories/polynesians-in-the-southern-ocean/>.

8 James Cook, *A Voyage Towards the South Pole, and Round the World, Performed in His Majesty's Ships the* Resolution *and* Adventure, *in the Years 1772, 1773, 1774, and 1775*, vol. 1, bk 1, December 1772, W Strahan & T Cadell, London, 1777.

9 James Clark Ross, *A Voyage of Discovery and Research in the Southern and Antarctic Regions, During the Years 1839–43*, vol. 2, J Murray, London, 1847, p. 210; 'Ross Ice Shelf', *Encyclopaedia Britannica*, 22 January 2016, accessed 13 February 2018, <www.britannica.com/place/Ross-Ice-Shelf>.

10 Roald Amundsen, *The South Pole: An Account of the Norwegian Antarctic Expedition in the* Fram, *1910–1912*, J Murray, London, 1912, p. 20.

11 Bernacchi, *To the South Polar Regions*, pp. 32, 31.

12 Information regarding and quotations from Joseph Matkin in this section are drawn from his *At Sea with the Scientifics: The* Challenger *Letters of Joseph Matkin*, ed. Philip F Rehbock, University of Hawaii Press, Honolulu, 1992, pp. 23, 139, 140, 145, 146–47.

13 Bernadette Hince, *The Antarctic Dictionary: A Complete Guide to Antarctic English*, CSIRO, Canberra, 2000, pp. 175–76.

14 Charles Wilkes's Termination Land was a large projection off the coast of Queen Mary Land, in Antarctica, which Douglas Mawson renamed Termination Ice Tongue during the Australasian Antarctic Expedition, in 1911–14. The ice tongue forms the eastern part of a larger ice shelf, which Mawson named Shackleton Ice Shelf, after Ernest Shackleton. See Australian Antarctic Division, 'Western Party: introduction', *Home of the Blizzard: The Australasian Antarctic Expedition*, last updated 3 July 2014, accessed 13 February 2018, <mawsonshuts.antarctica.gov.au/western-party>.

15 John Scott Keltie and Hugh Robert Mill (eds), *Report of the Sixth International Geographical Congress, Held in London, 1895*, J Murray, London, 1896, pp. 164, 176.

16 Clements R Markham, 'Address to the Royal Geographical Society', *Geographical Journal*, vol. 16, no. 1, July 1900, p. 13 (pp. 1–14).

17 The dedication in the front of the national library's *Antarctic Manual* reads, 'This book which formed part of the Library of the s.s. "Discovery" of the Antarctic Polar Expedition 1901–4, under Captain R.F. Scott, was presented to the Commonwealth Parliamentary Library by Sir R. Leicester Harmsworth, Bt., of London, England. September, 1931'. Only 500 copies of the book were printed. Nearly half of the *Discovery*'s library comprised books on matters of science. George Murray (ed.), *The Antarctic Manual for the Use of the Expedition of 1901*, Royal Geographical Society, London, 1901; 'Catalogue of the books of the "Discovery"', London, 1901, National Library of Scotland, Edinburgh, Selected polar ephemera, GB/A3844; Clements R Markham, *The Lands of Silence*, Cambridge University Press, Cambridge, 1921, p. 453.

18 Gordon Elliott Fogg, *A History of Antarctic Science*, Cambridge University Press, Cambridge, 1992, p. 116.

19 Edward J Larson, *An Empire of Ice: Scott, Shackleton, and the Heroic Age of Antarctic Science*, Yale University Press, New Haven, 2011, p. 252; John Francis Lovering, 'Gregory, John Walter (1864–1932)', *Australian Dictionary of Biography*, Australian National University, first published 1983, accessed 8 August 2017, <adb.anu.edu.au/biography/gregory-john-walter-6479>.

20 Edward Adrian Wilson, *Diary of the* Discovery *Expedition to the Antarctic Regions, 1901–1904*, ed. Ann Savours, Blandford, London, 1966, p. 74.

21 Pyne, *The Ice*, p. 56.

22 Wilson, *Diary of the* Discovery *Expedition*, p. 74.

23 Cited in Larson, *An Empire of Ice*, p. 253.

24 Pyne, *The Ice*, p. 7.

25 Wilson, *Diary of the* Discovery *Expedition*, p. 75.

26 Cited in Larson, *An Empire of Ice*, p. 255.

27 Earl Bathurst to Governor Ralph Darling, 'Despatch no. 16 per ship Manlius', 30 March 1827, in Commonwealth of Australia, *Historical Records of Australia*, series 1, Governor's Despatches to and from England, vol. 13, *January, 1827 – February, 1828*, ed. Frederick Watson, Library Committee of the Commonwealth Parliament, Sydney, 1920, p. 210.

28 The name 'krill' comes from a Norwegian word meaning 'small fry of fish'. Six species of krill are found in the Southern Ocean, and Antarctic krill (*Euphausia superba*) is the largest and most abundant.

29 'Katabatic' comes from the Greek *katabaino*, meaning 'to go down'. It is the generic term used to describe winds that flow down from mountains, plateaus and glaciers under the force of gravity.

30 Australian Antarctic Division, 'Emperor penguins', Department of the Environment and Energy (Australia), last updated 28 June 2017, accessed 7 August 2017, <antarctica.gov.au/about-antarctica/wildlife/animals/penguins/emperor-penguins>.

31 Bernard Stonehouse, 'Penguin biology: an overview', in Lloyd S Davis and John T Darby (eds), *Penguin Biology*, Academic, San Diego, 1990, pp. 6–7 (pp. 1–11).

32 Edward Adrian Wilson, 'Aves', in British National Antarctic Expedition, 1901–4, *Natural History*, vol. 2, *Zoology*, British Museum, London, 1907, p. 11.

33 Elizabeth Leane has made a study of Victorian perceptions of the Antarctic region. See, for example, Leane, 'Eggs, emperors and empire: Apsley Cherry-Garrard's "Worst Journey" as imperial quest romance', *Kunapipi*, vol. 31, no. 2, 2009, pp. 15–31.

34 Peter Dodson, 'Origin of birds: the final solution?', *American Zoologist*, vol. 40, no. 4, 2000, p. 504 (pp. 504–12).

35 Wilson, 'Aves', p. 31.

36 Diana Kwon, 'Embryonic evolution through Ernst Haeckel's eyes', *Scientist*, 1 May 2017, accessed 11 August 2017, <www.the-scientist.com/?articles.view/articleNo/49235/title/Embryonic-Evolution-Through-Ernst-Haeckel-s-Eyes/>.

37 Apsley Cherry-Garrard, *The Worst Journey in the World*, Vintage, London, 2010 (1922), pp. 274, 235.

38 Leane, 'Eggs, emperors and empire', p. 26.

39 New Zealand was used as a convenient resupply base for successive exploratory expeditions into the Southern Ocean and Antarctica, including the Russian explorer Fabian Gottlieb Thaddeus von Bellingshausen's voyage in 1819–20; Charles Wilkes's United States Exploring Expedition, in 1838–42; James Clark Ross's expedition to Antarctica, in 1839–43; the Norwegian explorer Carsten Borchgrevink's *Southern Cross* expedition, in 1898–1900; Robert Falcon Scott's *Discovery* expedition, in 1901–4, and *Terra Nova* expedition, in 1910–13; the Japanese explorer Nobu Shirase's expedition in 1910–12; the Australian explorer Douglas Mawson's expedition in 1911–14; and many subsequent twentieth-century scientific expeditions.

40 'The Nimrod's return: Lieutenant Shackleton's report; a wonderful voyage', *Albury Banner and Wodonga Express*, 13 March 1908, p. 7.

41 TW Edgeworth David distinguished himself during the expedition by leading the first ascent of Mount Erebus, the only active volcano on the Antarctic continent, accompanied by Alistair Mackay and David's former student Douglas Mawson, as expedition physicist. David Francis Branagan and Thomas George Vallance, 'David, Sir Tannatt William Edgeworth (1858–1934)', *Australian Dictionary of Biography*, Australian National University, first published 1981, accessed 8 August 2017, <adb.anu.edu.au/biography/david-sir-tannatt-william-edgeworth-5894>; David, 'With Shackleton. For the South Pole. British expedition, 1907. Notes by Professor David', pt 9, *Sydney Morning Herald*, 31 March 1908, p. 7; David, 'With Shackleton. For the South Pole. British expedition, 1907. Notes by Professor David', pt 8, *Sydney Morning Herald*, 30 March 1908, p. 7.

42 Information regarding and quotations from Frank Arthur Worsley to the end of this section are drawn from his Endurance*: An Epic of Polar Adventure*, P Allan, London, 1931, chart 3 (1916), pp. 45, 13, 44, 10. See also Ernest Shackleton, *South: The Story of Shackleton's Last Expedition, 1914–17*, Heinemann, London, 1919.

43 Bernard Stonehouse went on to become a renowned authority on king penguins

and seabirds in the Antarctic and subantarctic islands. Stonehouse, 'Shackleton's emperor penguins', *Polar Record*, vol. 50, no. 253, April 2014, p. 197 (pp. 192–98).

44 Barbara Wienecke, 'The history of discovery of emperor penguin colonies, 1902–2004', *Polar Record*, vol. 46, no. 3, July 2010, pp. 271–76.

45 International Union for Conservation of Nature, '*Aptenodytes forsteri*', *IUCN Red List of Threatened Species*, 2017, accessed 16 August 2017, <www.iucnredlist.org/details/22697752/0>; Woods Hole Oceanographic Institution, 'Finding new homes won't help emperor penguins cope with climate change', 7 June 2017, accessed 16 August 2017, <www.whoi.edu/news-release/emperor-penguin-dispersal>.

46 Bernacchi, *To the South Polar Regions*, p. 38.

47 Robert Falcon Scott, *Scott's Last Expedition: Diaries, 26 November 1910 – 29 March 1912*, Amberley, Stroud, 2012, pp. 58, 52–53.

48 Australian Antarctic Division, 'Adélie penguins', Department of the Environment and Energy (Australia), last updated 19 January 2017, accessed 10 August 2017, <antarctica.gov.au/about-antarctica/wildlife/animals/penguins/adelie-penguins>.

49 David, 'With Shackleton', pt 8.

50 Cited in Larson, *An Empire of Ice*, p. 180.

51 'Western Australian Museum specimens from the Antarctic', *Western Mail* (Perth), 29 January 1910, p. 28; 'Antarctic birds at the museum', *Advertiser* (Adelaide), 12 October 1909, p. 6.

52 Worsley, Endurance, pp. 44–45.

53 Norbert Wu and Jim Mastro, *Under Antarctic Ice: The Photographs of Norbert Wu*, University of California Press, Berkeley, 2004.

54 Scott, *Scott's Last Expedition*, pp. 53–54.

5 Deep

1 Rachel Carson, *The Sea Around Us*, Oxford University Press, Oxford, 1951, p. 75.

2 The first submarine telegraph cables were laid between 1857 and 1866. John Murray, '1899: the state of ocean science; an essay by Sir John Murray', *Scottish Geographical Magazine*, vol. 15, 1899, pp. 505–22.

3 Helen Rozwadowski, *Fathoming the Ocean: The Discovery and Exploration of the Deep Sea*, Belknap Press of Harvard University Press, Cambridge, 2009, pp. 6, 49–62.

4 For background and context to the Magnetic Crusade, see, for example, Edward J Larson, 'Public science for a global empire: the British quest for the South Magnetic Pole', *Isis*, vol. 102, no. 1, 2011, pp. 34–59; John Cawood, 'The Magnetic Crusade: science and politics in early Victorian Britain', *Isis*, vol. 70, no. 254, 1979, pp. 493–518.

5 Charles Wilkes, *Narrative of the United States Exploring Expedition During the Years 1838, 1839, 1840, 1841, 1842*, 5 vols, Lea & Blanchard, Philadelphia, 1845; Volker Siegel, 'Introducing Antarctic krill *Euphausia superba* Dana, 1850', in Volker Siegel (ed.), *Biology and Ecology of Antarctic Krill*, Springer, Switzerland, 2016, pp. 2–3 (pp. 1–20).

6 Gordon Elliott Fogg, *A History of Antarctic Science*, Cambridge University Press, Cambridge, 1992, pp. 55, 83.
7 See Thomas R Anderson and Tony Rice, 'Deserts on the sea floor: Edward Forbes and his azoic hypothesis for a lifeless deep ocean', *Endeavour*, vol. 30, no. 4, December 2006, pp. 131–37.
8 Edward Forbes, *The Natural History of the European Seas*, J Van Voorst, London, 1859, pp. 26–27.
9 Charles Wyville Thomson, 'Hydrographic instructions to Captain G. S. Nares, H.M.S. *Challenger*', in Thomas Henry Tizard, Henry Nottidge Moseley, John Young Buchanan and John Murray, *Narrative of the Cruise of H.M.S.* Challenger, *with a General Account of the Scientific Results of the Expedition*, vol. 1 of Great Britain, Challenger Office, *Report on the Scientific Results of the Voyage of H.M.S.* Challenger *During the Years 1873–76 under the Command of Captain George S. Nares, R.N., F.R.S. and the Late Captain Frank Tourle Thomson, R.N.*, Neill, Edinburgh, 1885, pt 1, p. 34 (pp. 34–39).
10 Charles Wyville Thomson, *The Depths of the Sea: An Account of the General Results of the Dredging Cruises of H.M.SS. 'Porcupine' and 'Lightning' During the Summers of 1868, 1869, and 1870, under the Scientific Direction of Dr. Carpenter, F.R.S., J. Gwyn Jeffreys, F.R.S., and Dr. Wyville Thomson, F.R.S.*, Macmillan, London, 1873, pp. 1–2, 4.
11 Charles Darwin, *On the Origin of Species by Means of Natural Selection; or, The Preservation of Favoured Races in the Struggle for Life*, J Murray, London, 1859; Charles Lyell, *Principles of Geology, Being an Attempt to Explain the Former Changes of the Earth's Surface, by Reference to Causes Now in Operation*, 3 vols, J Murray, London, 1830–3; Richard Corfield, *The Silent Landscape: The Scientific Voyage of HMS* Challenger, Joseph Henry, Washington, DC, 2003, pp. 5, 28.
12 Corfield, *The Silent Landscape*, p. 153.
13 Cited in Thomas Henry Huxley, 'The first volume of the publications of the "Challenger"', *Nature*, no. 23, November 1880, pp. 1–3.
14 Fogg, *A History of Antarctic Science*, pp. 95–96.
15 Herbert Swire, *The Voyage of the* Challenger*: A Personal Narrative of the Historic Circumnavigation of the Globe in the Years 1872–1876*, Golden Cockerel, London, 1938, p. 160.
16 Cited in Huxley, 'The first volume', p. 50.
17 Swire, *The Voyage of the* Challenger, p. 168.
18 John Murray, 'The renewal of Antarctic exploration', *Geographical Journal*, vol. 3, no. 1, January 1894, p. 21 (pp. 1–27); Corfield, *The Silent Landscape*, p. 153.
19 Murray, 'The renewal of Antarctic exploration', p. 19.
20 See Adriana Dutkiewicz, R Dietmar Muller, Simon O'Callaghan and Hjortur Jonasson, 'Census of seafloor sediments in the world's ocean', *Geology*, vol. 43, no. 9, 2015, pp. 795–98; Frank JR Taylor, 'Phytoplankton ecology before 1900', in Mary Sears and Daniel Merriman (eds), *Oceanography: The Past; Proceedings of the Third International Congress on the History of Oceanography*, Springer, New York, 2012, p. 512 (pp. 509–21).
21 Louis Bernacchi, *To the South Polar Regions: Expedition of 1898–1900*, Bluntisham, Denton, 1991 (1901), pp. 34–35.

22 Joseph Dalton Hooker, *The Botany of the Antarctic Voyage of H.M. Discovery Ships* Erebus *and* Terror *in the Years 1839–1843, under the Command of Captain Sir James Clark Ross, Kt., R.N., F.R.S. &c.*, vol. 1, *Flora Antarctica*, Reeve, London, 1844, p. 505; George A Knox, *Biology of the Southern Ocean*, 2nd edn, CRC, Boca Raton, 2007, p. 59.

23 See, for example, Thomas Henry Huxley, 'On some of the results of the expedition of H.M.S. *Challenger*', *Contemporary Review*, 1875, reproduced in Thomas Henry Huxley, *Collected Essays*, vol. 8, *Discourses: Biological and Geological*, D Appleton, New York, 1894, pp. 69–111.

24 Ernst Haeckel, *Kunstformen der Natur* [Art forms in nature], Bibliographisches Institut, Leipzig, 1904. See also Peter Williams, Dylan Evans, David Roberts and David Thomas, *Art Forms from the Abyss: Ernst Haeckel's Images from the HMS* Challenger *Expedition*, Prestel, New York, 2015.

25 See, for example, Howard Lynk, *Cabinet of Curiosities: A Selection of Antique Microscope Slides from the Victorian Era, c. 1830s–1900*, accessed 20 February 2018, <www.victorianmicroscopeslides.com/slides.htm>.

26 John Gwyn Jeffreys, 'Preliminary report of the biological results of a cruise in H.M.S. "Valorous" to Davis Strait in 1875', *Proceedings of the Royal Society of London*, vol. 25, no. 173, 1876, p. 186 (pp. 177–230).

27 John Murray completed the publication of the 50-volume *Challenger* report in 1895, following the death of Charles Wyville Thomson, in 1882, and is widely acknowledged as one of the founders of modern oceanography. Great Britain, Challenger Office, *Report on the Scientific Results of the Voyage of H.M.S.* Challenger *During the Years 1873–76 under the Command of Captain George S. Nares, R.N., F.R.S. and [the Late] Captain Frank Tourle Thomson, R.N.*, 50 vols, Neill, Edinburgh, 1880–95; Brian Price, 'Sir John Murray (1841–1914), founder of modern oceanography', University of Edinburgh, 1999, accessed 18 September 2017, <www.eeo.ed.ac.uk/public/JohnMurray.html>.

28 'Bathymetrical chart of the oceans showing the "deeps" according to Sir John Murray, 1899', NOAA Central Library Historical Collections, Silver Spring, Maryland, map00329.

29 Murray, 'The renewal of Antarctic exploration', pp. 1–27, 25. See Tizard et al., *Narrative of the Cruise of H.M.S.* Challenger, vol. 1. This report was condensed in John Murray and Johan Hjort, *Depths of the Ocean*, Macmillan, London, 1912, reflecting knowledge of deep-sea fauna at the end of the era of great nineteenth-century voyages of ocean exploration.

30 Karl Weyprecht, 'Scientific work of the second Austro-Hungarian polar expedition, 1872–4', *Journal of the Royal Geographical Society of London*, no. 45, 1875, pp. 19–33, cited in Fogg, *A History of Antarctic Science*, pp. 103–104.

31 An early use of the term 'heroic age' in relation to Antarctic exploration appeared in Margery Fisher and James Fisher, *Shackleton*, J Barrie, London, 1957, and was repeated in subsequent reviews, such as HJM, 'Vital biography of "Old Gruffy"', *Canberra Times*, 11 July 1959, p. 11. See also Tom Griffiths, *Slicing the Silence: Voyaging to Antarctica*, NewSouth, Sydney, 2007, p. 11; Bernadette Hince, *The*

Antarctic Dictionary: A Complete Guide to Antarctic English, CSIRO, Canberra, 2000, p. 167.

32 Ivar Hamre, 'The Japanese south polar expedition of 1911–12: a little-known episode in Antarctic exploration', *Geographical Journal*, vol. 82, no. 5, November 1933, pp. 411–23; Stephanie Pain, 'Scott, Amundsen … and Nobu Shirase', *New Scientist*, 20 December 2011, accessed 29 December 2017, <www.newscientist.com/article/mg21228440-900-scott-amundsen-and-nobu-shirase/>; Mariska M Wouters, 'One man's dream: Japan's Antarctic history', Graduate Certificate thesis, University of Canterbury, Christchurch, 1999, pp. 5–8.

33 Fogg, *A History of Antarctic Science*, p. 196.

34 Narissa Bax (Institute of Marine and Antarctic Studies, University of Tasmania), pers. comm., 8 December 2017.

35 Cited in 'The ocean deep', *Shepparton Advertiser*, 3 December 1935, p. 9.

36 William Beebe, *The World Beneath the Sea*, Crowell, New York, 1953.

37 Joe Hlebica, 'Roger Revelle and the great age of exploration', *Explorations*, vol. 8, no. 1, summer 2001, pp. 22–29.

38 'Scientists will explore ocean bed for sea monsters', *Mercury* (Hobart), 7 March 1950, p. 3. For the expedition's findings, see Anton F Bruun, Svend Greve, Hakon Mielche and Ragnar Spärck (eds), *The* Galathea *Deep Sea Expedition, 1950–1952, Described by Members of the Expedition*, trans. Reginald Spink, G Allen & Unwin, London, 1956.

39 'Scientists will explore ocean bed'.

40 Jules Verne, *Vingt mille lieues sous les mers: tour du monde sous-marin* [Twenty Thousand Leagues Under the Sea: A Tour of the Underwater World], Hetzel, Paris, 1871 (serialised 1869–70).

41 Anthonie Cornelis Oudemans, *The Great Sea Serpent: An Historical and Critical Treatise, with the Reports of 187 Appearances (Including Those of the Appendix), the Suppositions and Suggestions of Scientific and Non-scientific Persons, and the Author's Conclusions*, EJ Brill, Leiden, 1892. See also Franziska Torma, 'Snakey waters; or, How marine biology structured global environmental sciences', in Agnes Kneitz and Marc Landry (eds), 'On water: perceptions, politics, perils', *RCC Perspectives*, 2012/2, pp. 13–21.

42 'Sea monster on the coast of Tasmania', *Daily Telegraph* (Launceston), 26 May 1913, p. 5; Torma, 'Snakey waters', p. 14.

43 Michel J Kaiser, *Marine Ecology: Processes, Systems, and Impacts*, Oxford University Press, Oxford, 2011, p. 79.

44 The deepest trawls were conducted in the Philippine Trench. Anton F Bruun, 'Animal life of the deep sea bottom', in Bruun et al., *The* Galathea *Deep Sea Expedition*, pp. 179–80, 189 (pp. 149–95); Claude E ZoBell and Richard Y Morita, 'Bacteria in the deep sea', in Bruun et al., *The* Galathea *Deep Sea Expedition*, p. 204 (pp. 202–10).

45 Tim Winton, *Land's Edge: A Coastal Memoir*, Hamish Hamilton, Melbourne, 2010, pp. 33–34.

46 George ER Deacon (ed.), *Oceans: An Atlas-history of Man's Exploration of the Deep*, P Hamlyn, London, 1962, p. 207.

47 Jacques Cousteau, 'The perils and potentials of a watery planet', *Saturday Review*, 3 April 1974, pp. 41–45.

48 For example, see Alister C Hardy, *The Open Sea: Its Natural History*, vol. 1, *The World of Plankton*, Houghton Mifflin, Boston, 1956; Klaus Günther and Kurt Deckert, *Creatures of the Deep Sea*, trans. EW Dickes, Scribner's, New York, 1956; Clarence P Idyll, *Abyss: The Deep Sea and the Creatures That Live in It*, Crowell, New York, 1965.

49 Henry M Stommel, *Science of the Seven Seas*, Cornell Maritime Press, New York, 1945, pp. 29, 30–31.

50 Woods Hole Oceanographic Institution, 'About Henry Melson Stommel', accessed 26 July 2017, <www.whoi.edu/page.do?pid=7716>.

51 Matthew Fontaine Maury, *The Physical Geography of the Sea*, 8th edn, Harper, New York, 1860, p. xiv; Robert Kunzig, *Mapping the Deep: The Extraordinary Story of Ocean Science*, WW Norton, New York, 2000, p. 289; Trent Knoss, 'Mapping the storms of the sea', *MIT Technology Review*, 20 February 2013, accessed 19 January 2018, <www.technologyreview.com/s/510856/mapping-the-storms-of-the-sea/>; Charles D Hollister, Arthur RM Nowell and Peter A Jumars, 'The dynamic abyss', *Scientific American*, vol. 250, no. 3, March 1984, pp. 42–53.

52 Carson, *The Sea Around Us*, p. 14.

53 Amanda Hagood, 'Wonders with the sea: Rachel Carson's ecological aesthetic and the mid-century reader', *Environmental Humanities*, vol. 2, May 2013, pp. 57–77.

54 Carson, *The Sea Around Us*, p. 147.

55 The 'twilight' zone is formally known as the 'dysphotic' zone, where the amount of light able to penetrate rapidly diminishes. Below 1000 metres lies the 'midnight', or 'aphotic', zone, where there is complete darkness. SJ Baker, 'Rare studies of the sea', *Sydney Morning Herald*, 2 August 1952, p. 7.

56 'Three musts on your bookshelf', *Australian Women's Weekly*, 3 June 1959, p. 34; Baker, 'Rare studies of the sea'; 'Fascinating tale in *Daily Mirror*', *Truth* (Sydney), 10 August 1952, p. 2.

57 Maury, *The Physical Geography of the Sea*, 8th edn, p. 251.

58 William Thomas Blanford, 'Geology', in George Murray (ed.), *The Antarctic Manual for the Use of the Expedition of 1901*, Royal Geographical Society, London, 1901, p. 178 (pp. 176–87).

59 Eduard Suess, *Das Antlitz der Erde*, 3 vols, F Tempsky, Vienna, 1885–1901 (published as *The Face of the Earth*, trans. WJ Sollas, 5 vols, Clarendon, Oxford, 1904–24).

60 Robert Falcon Scott, *Scott's Last Expedition*, vol. 1, *Being the Journals of Captain R. F. Scott, R.N., C.V.O.*, 3rd edn, ed. Leonard Huxley, Smith, Elder, London, 1913, p. 564.

61 Australian Antarctic Division, 'Antarctic prehistory', Department of the Environment and Energy (Australia), last updated 14 June 2002, accessed 29 August 2017, <antarctica.gov.au/about-antarctica/environment/geology/antarctic-prehistory>.

62 Alfred Wegener, *Die Entstehung der Kontinente und Ozeane* [The origin of continents and oceans], Vieweg, Braunschweig, 1915.

63 Alfred Wegener's thesis, published in 1915, was translated into English in 1922 but failed to attract much interest internationally before the onset of wartime hostilities. Those who supported the theory included South African geologist Alexander Du Toit, in *Our Wandering Continents: An Hypothesis of Continental Drifting*, Oliver & Boyd, Edinburgh, 1937.

64 Marie Tharp, 'Connect the dots: mapping the seafloor and discovering the mid-ocean ridge', in Laurence Lippsett (ed.), *Lamont-Doherty Earth Observatory of Columbia: Twelve Perspectives on the First Fifty Years, 1949–1999*, Palisades, New York, 1999, pp. 31–37.

65 American Museum of Natural History, 'Harry Hess: one of the discoverers of seafloor spreading', accessed 19 January 2018, <www.amnh.org/explore/resource-collections/earth-inside-and-out/harry-hess-one-of-the-discoverers-of-seafloor-spreading/>, excerpt from Edmond A Mathez (ed.), *Earth: Inside and Out*, American Museum of Natural History / New Press, New York, 2000, p. 113.

66 Knox, *Biology of the Southern Ocean*, pp. 3–4; Tharp, 'Connect the dots'.

67 The text to the end of this chapter draws on and quotes from Bruce C Heezen and Charles D Hollister, *The Face of the Deep*, Oxford University Press, New York, 1971, pp. v, 3, 9, 19, 605, 257, 335, 372. Helen Rozwadowski explores aspects of the deep sea as frontier in *Fathoming the Ocean*, p. 29, and 'Arthur C Clarke and the limitations of the ocean as frontier', *Environmental History*, no. 17, 2012, pp. 578–602.

68 See Heezen and Hollister, *The Face of the Deep*, p. 261, fig. 7.31.

6 Current

1 Henry Beston, *The Outermost House: A Year of Life on the Great Beach of Cape Cod*, Doubleday, New York, 1928, p. 25.

2 British Antarctic Survey, 'Most biologically rich island in Southern Ocean', *Science Daily*, 31 May 2011, accessed 24 November 2017, <www.sciencedaily.com/releases/2011/05/110525181414.htm>.

3 Robert Burton, *South Georgia*, 4th edn, Government of South Georgia and the South Sandwich Islands, Stanley, 2016, pp. 18–19, 42.

4 'Bottom water', *Encyclopaedia Britannica*, 25 May 2017, accessed 25 November 2017, <www.britannica.com/science/bottom-water>.

5 Ryan Smith, Melicie Desflots, Sean White, Arthur J Mariano and Edward H Ryan, 'The Antarctic CP Current', Rosenstiel School of Marine and Atmospheric Science, University of Miami, 2013, accessed 24 June 2017, <oceancurrents.rsmas.miami.edu/southern/antarctic-cp.html>.

6 Stephen Rintoul and John Church, 'The Southern Ocean's global reach', *Australian Antarctic Magazine*, no. 4, spring 2002, pp. 2–4.

7 Professor Mike Coffin, pers. comm., 29 September 2017. See also Catherine L Schuur, Millard F Coffin, Cliff Frohlich, Christina G Massell, Garry D Karner, Doug Ramsay and David W Caress, 'Sedimentary regimes at the Macquarie Ridge Complex: interaction of Southern Ocean circulation and plate boundary

bathymetry', *Paleooceanography*, vol. 13, no. 6, December 1998, pp. 646–70.

8 See, for example, Howie D Scher and Ellen E Martin, 'Timing and climatic consequences of the opening of Drake Passage', *Science*, vol. 312, no. 5772, 21 April 2006, pp. 428–30; Howie D Scher, Joanne M Whittaker, Simon E Williams, Jennifer C Latimer, Wendy EC Kordesch and Margaret L Delany, 'Onset of Antarctic Circumpolar Current 30 million years ago as Tasmanian Gateway aligned with westerlies', *Nature*, vol. 523, no. 7562, 30 July 2015, pp. 580–83.

9 George A Knox, *Biology of the Southern Ocean*, 2nd edn, CRC, Boca Raton, 2007, p. 2.

10 Roy Livermore, 'Opening of Drake Passage', in Beau Riffenburgh (ed.), *Encyclopedia of the Antarctic*, Taylor & Francis, Didcot, 2007, pp. 344–46.

11 Rintoul and Church, 'The Southern Ocean's global reach'.

12 Australian Antarctic Division, 'Ocean fronts and circulation', Department of the Environment and Energy (Australia), last modified 28 February 2005, accessed 26 October 2017, <heardisland.antarctica.gov.au/nature/marine-life/ocean-fronts-and-circulation>.

13 James Cook, *A Voyage Towards the South Pole, and Round the World, Performed in His Majesty's Ships the* Resolution *and* Adventure, *in the Years 1772, 1773, 1774, and 1775*, vol. 1, bk 1, December 1772, W Strahan & T Cadell, London, 1777.

14 *Aurora* (the official journal of the ANARE Club), December 1974, p. 43.

15 For an animation of Southern Ocean circulation, see Research School of Earth Sciences, 'Big data reveals glorious animation of Antarctic Bottom Water', Australian National University, accessed 5 October 2017, <rses.anu.edu.au/research/research-stories/big-data-reveals-glorious-animation-antarctic-bottom-water>.

16 Eric H Christiansen and W Kenneth Hamblin, *Dynamic Earth: An Introduction to Physical Geology*, Jones & Bartlett Learning, Burlington, 2015, pp. 250–51.

17 Sandra Connelly, 'Antarctic Circumpolar Current', in S George Philander (ed.), *Encyclopedia of Global Warming and Climate Change*, vol. 1, Sage, California, 2012, p. 60 (pp. 60–61); Eric L Mills, *The Fluid Envelope of Our Planet: How the Study of Ocean Currents Became a Science*, University of Toronto, Toronto, 2009, p. 4.

18 See Edmond Halley, 'Halley's chart of the trade winds and monsoons, taken from *Bibliothèque universelle et historique de l'année, 1687, tome quatrième*', 1687, State Library of New South Wales, Sydney, RB/2785.

19 Gordon Elliott Fogg, *A History of Antarctic Science*, Cambridge University Press, Cambridge, 1992, pp. 13–15.

20 Joseph L Reid, 'Deep ocean circulation', in Richard C Vetter (ed.), *Oceanography: The Last Frontier*, Basil, New York, 1973, p. 207 (pp. 203–15).

21 Benjamin Graf von Rumford, *Collected Works of Count Rumford: The Nature of Heat*, Harvard University Press, Cambridge, 1968, p. 209 (original emphasis).

22 Mills, *The Fluid Envelope of Our Planet*, p. 5.

23 John Cawte Beaglehole, *The Journals of Captain James Cook on his Voyages of Discovery*, vol. 2, *The Voyage of the* Resolution *and* Adventure, *1772–1775*, Cambridge University Press for the Hakluyt Society, Cambridge, 1961, pp. 725, 68 n. 2.

24 Fogg, *A History of Antarctic Science*, p. 24.
25 George ER Deacon, *The Antarctic Circumpolar Ocean*, Cambridge University Press, Cambridge, 1984, p. 22.
26 Fabian Gottlieb Thaddeus von Bellingshausen, *The Voyage of Captain Bellingshausen to the Antarctic Seas, 1819–1821*, vol. 1, ed. Frank Debenham, trans. from Russian, Hakluyt Society, London, 1945, p. 28; Constantine M Hotimsky, 'Bellingshausen, Faddei Faddeevich (Fabian) (1778–1852)', *Australian Dictionary of Biography*, Australian National University, first published 1966, accessed 29 December 2017, <adb.anu.edu.au/biography/bellingshausen-faddei-faddeevich-fabian-1767>.
27 Matthew Fontaine Maury, 'Remarks on the Gulf Stream and currents of the sea', *American Journal of Science*, vol. 47, 1844, p. 181 (pp. 161–81).
28 Matthew Fontaine Maury, *The Physical Geography of the Sea*, 3rd edn, Harper, New York, 1855, pp. 138, 125, xiv.
29 Michael S Reidy, *Tides of History: Ocean Science and Her Majesty's Navy*, University of Chicago Press, Chicago, 2009, p. 37.
30 John Walter Gregory, *The Climate of Australasia in Reference to Its Control by the Southern Ocean*, Whitcombe & Tombs, Melbourne, 1904, pp. 39–40.
31 Gregory, *The Climate of Australasia*, pp. 30, 41.
32 'The week', *World's News* (Sydney), 23 January 1904, pp. 14–15.
33 'Weather prophecy: how it is done', *Argus* (Melbourne), 18 January 1908, p. 6.
34 These expeditions included the *Belgica* (1897–9), *Valdivia* (1898–9), *Gauss* (1901–3), *Antarctic* (1901–4), *Scotia* (1902–4), *Terra Nova* (1910–13), *Deutschland* (1911–12), *Aurora* (1911–14) and *Endurance* (1914–17). See Fogg, *A History of Antarctic Science*, pp. 196–201.
35 Fogg, *A History of Antarctic Science*, pp. 204, 197–98.
36 Adele Morrison, Thomas Frölicher and Jorge Sarmiento, 'Upwelling in the Southern Ocean', *Physics Today*, vol. 68, no. 1, 2015, p. 27 (pp. 27–32).
37 Knox, *Biology of the Southern Ocean*, pp. 355–57; Isobel Bennett, *Shores of Macquarie Island*, Rigby, Adelaide, 1971, p. 39.
38 Rachel Carson, *The Sea Around Us*, Oxford University Press, Oxford, 1951, p. 138.
39 Knox, *Biology of the Southern Ocean*, p. 232.
40 George L Small, *The Blue Whale*, Columbia University Press, New York, 1971, p. 45.
41 Tom Griffiths, *Slicing the Silence: Voyaging to Antarctica*, NewSouth, Sydney, 2007, p. 55; Alister C Hardy, *Great Waters: A Voyage of Natural History*, Harper & Row, New York, 1967, p. 34; John P Croxall and Phil N Trathan, 'The Southern Ocean: a model system for conserving resources', in Linda K Glover and Sylvia Earle (eds), *Defying Ocean's End: An Agenda for Action*, Island, Washington, DC, 2004, p. 73 (pp. 71–88).
42 Archaeological surveys reveal the extent of shore-based whaling (and sealing) sites along these southern coasts and islands. See, for example, Susan Lawrence and Mark Staniforth (eds), *The Archaeology of Whaling in Southern Australia and New Zealand*, Brolga Press for Australasian Society for Historical Archaeology and Australian Institute for Maritime Archaeology, Gundaroo, 1998.
43 John Newton notes at least 35 in *Savage History*, NewSouth, Sydney, 2013, p. 125, but Michael Nash identifies 59, based on previous historical and archaeological

research, in *The Bay Whalers: Tasmania's Shore-based Whaling Industry*, Navarine, Woden, 2003, pp. 127–60.

44 Lorne K Kriwoken and John W Williamson, 'Hobart, Tasmania: Antarctic and Southern Ocean connections', *Polar Record*, vol. 29, no. 169, April 1993, pp. 93–102; Bjørn L Basberg and Robert K Headland, 'The economic significance of the 19th century Antarctic sealing industry', *Polar Record*, vol. 49, no. 4, October 2013, pp. 381–91.

45 The British Government changed the name Van Diemen's Land to Tasmania in 1856. WH Leigh, *Reconnoitering Voyages and Travels with Adventures in the New Colonies of South Australia*, Smith, Elder, London, 1840, pp. 169–70.

46 The Portland whaling industry declined from the early 1840s as wool-growing became more profitable. John Stanley Cumpston, 'Dutton, William (1811–78)', *Australian Dictionary of Biography*, Australian National University, first published 1966, accessed 17 January 2018, <adb.anu.edu.au/biography/dutton-william-2011>.

47 Cited in SMS, 'Whaling was once a stimulus to settlement', *Advocate* (Burnie), 19 September 1953, p. 10.

48 Lynette Russell's study gives a detailed picture of the relationship between Aboriginal Australian coastal people and white whalers and sealers in the period. Russell, *Roving Mariners: Australian Aboriginal Whalers and Sealers in the Southern Oceans, 1790–1870*, State University of New York Press, Albany, 2012, pp. 32, 41, 29, 24–26.

49 Joseph Lycett, *Aborigines Cooking and Eating Beached Whales, Newcastle, New South Wales, ca. 1817*, c. 1817, watercolour, 17.7 × 27.9 cm, National Library of Australia, Canberra, PIC MSR 12/1/4 #R5680.

50 Russell, *Roving Mariners*, p. 26.

51 Cited in 'Early days: a valuable diary', *Journal* (Adelaide), 24 August 1918, p. 6.

52 James Clark Ross, *A Voyage of Discovery and Research in the Southern and Antarctic Regions, During the Years 1839–43*, vol. 1, J Murray, London, 1847, pp. 191–92.

53 Charles Enderby, *The Auckland Islands: A Short Account of Their Climate, Soil, & Productions: And the Advantages of Establishing There a Settlement at Port Ross for Carrying on the Southern Whale Fisheries*, P Richardson, London, 1849. See also Marilyn Landis, *Antarctica: Exploring the Extreme*, Chicago Review Press, Chicago, 2001, p. 111; Kenneth M Dallas, 'Enderby, Samuel (1756–1829)', *Australian Dictionary of Biography*, Australian National University, first published 1966, accessed 17 January 2018, <adb.anu.edu.au/biography/enderby-samuel-2026>.

54 Fergus B McLaren, *The Auckland Islands: Their Eventful History*, AH & AW Reed, Wellington, 1948, p. 57, cited in Bernadette Hince, 'The teeth of the wind: an environmental history of subantarctic islands', unpublished PhD thesis, Australian National University, Canberra, 2005, p. 197.

55 Department of Conservation, Te Papa Atawhai (New Zealand), 'Auckland Islands', accessed 17 October 2017, <www.doc.govt.nz/parks-and-recreation/places-to-go/southland/places/subantarctic-islands/auckland-islands/>.

56 The Ngāi Tahu people, of the South Island of New Zealand, have an oral

history of fishing and hunting in the surrounding seas and claim *manawhenua* (customary rights) for the Aucklands. See Department of Conservation, Te Papa Atawhai (New Zealand), 'Auckland Islands: Motu Maha Marine Reserve', accessed 17 October 2017, <www.doc.govt.nz/parks-and-recreation/places-to-go/southland/places/subantarctic-islands/auckland-islands/auckland-islands-motu-maha-marine-reserve/>.

57 The islands were officially incorporated as New Zealand territory in 1863. The whales were apparently fished out by the time the Enderby settlement was established. See Hince, 'The teeth of the wind', pp. 213–14.

58 Michael J Moore, Simon D Berrow, Brenda A Jensen, Pauline Carr, Richard Sears, Victoria J Rowntree, Roger Payne and Philip K Hamilton, 'Relative abundance of large whales around South Georgia (1979–1998)', *Marine Mammal Science*, vol. 15, no. 4, October 1999, p. 1288 (pp. 1287–1302).

59 J Gunnar Andersson, 'The winter expedition of the *Antarctic* to South Georgia', *Geographical Journal*, vol. 20, no. 4, October 1902, pp. 405–408.

60 Carl Anton Larsen collection, Scott Polar Research Institute Archives, University of Cambridge, Cambridge.

61 Robert Cushman Murphy, *Logbook for Grace: Whaling Brig Daily, 1912–1913*, R Hale, London, 1948, p. 143.

62 Francis Downes Ommanney, *Lost Leviathan: Whales and Whaling*, Dodd, Mead, New York, 1971, p. 131.

63 Murphy, *Logbook for Grace*, p. 150.

64 Department of Commerce and Agriculture (Australia), *Japanese Antarctic Whaling Operations, 1946/47: Report on the Operations of the Japanese Mother Ship 'Hashidate Maru' and Attendant Chasers, Based on Information Supplied by the Official Australian Observer*, Sydney, May 1947, p. 51.

65 Murphy, *Logbook for Grace*, pp. 150, 142.

66 Robert J Hofman, 'Sealing, whaling and krill fishing in the Southern Ocean: past and possible future effects on catch regulations', *Polar Record*, vol. 53, no. 1, January 2017, p. 90 (pp. 88–99).

67 Fogg, *A History of Antarctic Science*, pp. 155–56.

68 'Southern Ocean', *Eastern Districts Chronicle* (York, Western Australia), 17 July 1914, p. 7. See also Philip Hoare, *Leviathan; or, The Whale*, Fourth Estate, London, 2008, pp. 317–18.

69 See Lucinda Moore, *Animals in the Great War*, Pen & Sword, Barnsley, 2017.

70 Hoare, *Leviathan*, p. 316; Hardy, *Great Waters*, pp. 36–37; Fogg, *A History of Antarctic Science*, pp. 155–56.

71 Cited in Hoare, *Leviathan*, p. 316.

72 Bennett, *Shores of Macquarie Island*, p. 14. See also Small, *The Blue Whale*; Hofman, 'Sealing, whaling and krill fishing in the Southern Ocean', p. 90.

73 EW Hunter Christie, *The Antarctic Problem: An Historical and Political Study*, G Allen & Unwin, London, 1951; Fogg, *A History of Antarctic Science*, p. 155; Rodney Russ, 'History, exploration, settlement and past use of the sub-antarctic', *Papers and Proceedings of the Royal Society of Tasmania*, vol. 141, no. 1, 2007, pp. 169–72.

74 Alister C Hardy designed the continuous plankton recorder during the voyage and pioneered the study of the distribution and abundance of plankton in the ocean. Hardy, *Great Waters*, p. 20. See also Alister C Hardy and Eugene R Gunther, 'The plankton of the South Georgia whaling grounds and adjacent waters, 1926–7', in Discovery Committee, Colonial Office, *Discovery Reports*, vol. 11, Cambridge University Press, Cambridge, 1936, pp. 1–146.

75 The site is now used by the British Antarctic Survey for fisheries research. Fogg, *A History of Antarctic Science*, p. 211.

76 George Deacon noted that by 1975 a total of 8234 whales had been marked by *Discovery* scientists and subsequent marking projects, of which only 10.5 per cent were ever recovered. Knox, *Biology of the Southern Ocean*, p. 211; Deacon, *The Antarctic Circumpolar Ocean*, p. 65. Books published by scientists involved in the Discovery Investigations include Francis Downes Ommanney, *South Latitude*, Longmans, Green, London, 1938; John Coleman-Cooke, Discovery II *in the Antarctic*, Odhams, London, 1963; and Hardy, *Great Waters*.

77 The British, Australian and New Zealand Antarctic Research Expedition team mapped the Antarctic coastline, and this work was eventually used to support Australia's case for establishing the Australian Antarctic Territory, in 1936. 'Antarctica. Abundance of whalers. Amazing mirage. Cruise of *Discovery*', *Sydney Morning Herald*, 11 May 1931, p. 10.

78 Bjørn L Basberg, 'Larsen, Carl Anton', in Riffenburgh, *Encyclopedia of the Antarctic*, pp. 584–85.

79 Gilbert Eric Douglas accompanied the expedition as one of two air force pilots to undertake reconnaissance flights and guide the *Discovery* through the sea ice. Douglas, 'Whaling factory ship: Eric Douglas; December, 1930', BANZARE log, 15, 16 December 1930, transcript, *Trove*, list created by 'beetle', 30 December 2011, accessed 2 November 2017, <www.trove.nla.gov.au/list?id=18812>; David Wilson, 'Douglas, Gilbert Eric (1902–1970)', *Australian Dictionary of Biography*, Australian National University, first published 1996, accessed 17 January 2018, <adb.anu.edu.au/biography/douglas-gilbert-eric-10038>.

80 Adam Nicolson, 'The rise', *Britain's Whale Hunters: The Untold Story*, television series, BBC, July 2017.

81 Ommanney, *Lost Leviathan*, p. 145.

82 Fogg, *A History of Antarctic Science*, pp. 240, 387.

83 *Convention for the Regulation of Whaling*, Geneva, 24 September 1931, 155 LNTS 349, entered into force 16 January 1935. See also Hoare, *Leviathan*, pp. 320–21; Knox, *Biology of the Southern Ocean*, pp. 211–13.

84 By 2017 the International Whaling Commission had 89 members, of which three – Japan, Norway and Iceland – were engaged in commercial whaling. *International Convention for the Regulation of Whaling*, Washington, 2 December 1946, 161 UNTS 72, entered into force 10 November 1948, p. 1.

85 Moore et al., 'Relative abundance of large whales', p. 1296.

86 Micheline Jenner, *The Secret Life of Whales*, NewSouth, Sydney, 2017, pp. 241–42.

87 Integrated Marine Observing System, 'Continental shelf and coastal processes',

accessed 17 October 2017, <www.imos.org.au/nodes/saimos/saimossciback/saimosscibackcsp/>.

88 Cited in Tony Wright, 'The whale song man, the oil giant and the Great Australian Bight', *Sydney Morning Herald*, 15 October 2015, accessed 5 May 2017, <www.smh.com.au/federal-politics/political-opinion/the-whale-song-man-the-oil-giant-and-the-great-australian-bight-20151015-gk9qwm.html>.

89 Cited in Sam Burcher, 'The return of the whale dreamers', *Science in Society Archive*, 9 September 2009, accessed 5 May 2017, <www.i-sis.org.uk/WhaleDreamers.php>.

90 Kim Kindersley and Julian Lennon collaborated to make 'Eyes of the soul: legends of whales, dolphins and tribes', but it was not completed. See CJ Burianek, 'Eyes of the soul', *Hey Jules*, updated 2008, accessed 18 February 2018, <www.heyjules.com/wouldyou/eyesofthesoul.html>.

91 Kim Kindersley, *Whaledreamers*, documentary film, Heart Magic Media and Youngheart Entertainment, Australia, 2006.

92 The British Government also carried out nuclear tests on the Montebello Islands, off the northwestern coast of Western Australia. National Archives of Australia, 'British nuclear tests at Maralinga', Fact Sheet 129, accessed 5 May 2017, <www.naa.gov.au/collection/fact-sheets/fs129.aspx>.

93 Goldfields Land and Sea Council, 'Mirning people: native title determination', media statement, 25 October 2017, accessed 9 December 2017, <www.glsc.com.au/media-statements/mirningpeople-nativetitledetermination>.

94 Department of the Environment and Energy (Australia), 'Great Australian Bight Commonwealth Marine Reserve', accessed 10 December 2017, <www.environment.gov.au/topics/marine/marine-reserves/south-west/gab>.

95 The Senate inquiry lapsed when the Australian Government called a federal election for July 2016. Senate Environment and Communications References Committee (Australia), 'Oil or gas production in the Great Australian Bight', Parliament of Australia, accessed 30 April 2017, <www.aph.gov.au/Parliamentary_Business/Committees/Senate/Environment_and_Communications/Oil_drill_Great_Aus_Bight>.

96 Sea Shepherd Australia, 'Conservationist Bob Brown and actor David Field join the *Steve Irwin* vessel in the fight for the bight', 15 August 2016, accessed 30 June 2017, <www.seashepherd.org.au/news-and-commentary/news/conservationist-bob-brown-and-actor-david-field-join-the-steve-irwin-vessel-in-the-fight-for-the-bight.html>; 'BP withdraws from Great Australian Bight drilling', *ABC News*, 11 October 2016, accessed 30 June 2017, <www.abc.net.au/news/2016-10-11/bp-withdraws-from-great-australian-bight-drilling/7921956>; 'Great Australian Bight oil rigs back on the drawingboard', *New Daily*, 9 June 2017, accessed 14 February 2018, <www.thenewdaily.com.au/news/state/sa/2017/06/09/oil-gas-company-drill-bight/>.

7 Convergence

1 TS Eliot, 'The dry salvages', in *Four Quartets*, Harcourt, San Diego, 1941, p. 38 (pp. 35–48).

2 Herbert Ponting, *The Great White South; or, With Scott in the Antarctic, Being an Account of Experiences with Captain Scott's South Pole Expedition and of the Natural Life of the Antarctic*, Duckworth, London, 1935 (1921), p. 15.
3 Rachel Carson, *The Sea Around Us*, Oxford University Press, Oxford, 1951, p. xxi.
4 Robert C Cowen, *Frontiers of the Sea: The Story of Oceanographic Exploration*, V Gollancz, London, 1960, p. 20; Gardner Soule, *The Ocean Adventure: Science Explores the Depths of the Sea*, Appleton-Century, New York, 1966, p. 5.
5 Charles C Bates, Thomas F Gaskell and Robert B Rice, *Geophysics in the Affairs of Man: A Personalized History of Exploration Geophysics and Its Allied Sciences of Seismology and Oceanography*, Elsevier, Amsterdam, 2016, p. 152.
6 WJ Dakin, 'The whale has no privacy', *Sunday Times* (Perth), 22 June 1947, p. 1.
7 *The Antarctic Treaty*, Washington, DC, 1 December 1959, 402 UNTS 71, entered into force 23 June 1961.
8 Bates, Gaskell and Rice, *Geophysics in the Affairs of Man*, p. 159.
9 Sayed Z El-Sayed (ed.), *Southern Ocean Ecology: The BIOMASS Perspective*, Cambridge University Press, Cambridge, 1994; Sayed Z El-Sayed, 'Understanding the Antarctic marine ecosystem: a prerequisite for its conservation', in Bruce C Parker (ed.), *Proceedings of the Colloquium on Conservation Problems in Antarctica*, Virginia Polytechnic Institute and State University, Blacksburg, 1972, p. 136 (pp. 131–42); Alessandro Antonello, 'The greening of Antarctica: environment, science and diplomacy, 1959–1980', PhD thesis, Australian National University, Canberra, March 2014, pp. 205–206.
10 Isobel Bennett, *Shores of Macquarie Island*, Rigby, Adelaide, 1971, pp. 15–16.
11 Bruce McLennan, *The History of Oceans Governance*, Monograph Series no. 9, Australian Defence College, Canberra, 2006, p. 1.
12 Hugo Grotius, *Mare liberum* [The freedom of the seas], L Elzevir, Leiden, 1609.
13 McLennan, *The History of Oceans Governance*, pp. 17–22.
14 Micheline Jenner, *The Secret Life of Whales*, NewSouth, Sydney, 2017, p. 243.
15 Michael J Moore, Simon D Berrow, Brenda A Jensen, Pauline Carr, Richard Sears, Victoria J Rowntree, Roger Payne and Philip K Hamilton, 'Relative abundance of large whales around South Georgia (1979–1998)', *Marine Mammal Science*, vol. 15, no. 4, October 1999, p. 1289 (pp. 1287–1302); Robert C Rocha Jr, Phillip J Clapham and Yulia V Ivashchenko, 'Emptying the oceans: a summary of industrial whaling catches in the 20th century', *Marine Fisheries Review*, vol. 76, no. 4, 2014, pp. 37–48.
16 See, for example, Robert Hunter, *To Save a Whale: The Voyages of Greenpeace*, Heinemann, London, 1978.
17 'The last whale', *Hindsight*, ABC Radio National, 2 June 2013, accessed 6 December 2017, <www.abc.net.au/radionational/programs/hindsight/hindsight-sunday-2-june-2013/4711342>.
18 Under international law, countries with coastlines have the right to establish an economic exclusion zone extending 370 kilometres from shore. The United States introduced the world's first economic exclusion zone, in 1945, to control the natural resources of its continental shelf, and other nations followed suit. Australia declared its zone in 1994.

19 James D Hansom and John E Gordon, *Antarctic Environments and Resources: A Geographical Perspective*, Routledge, Abingdon, 2013 (1998), p. 205.
20 See Kjellrun Hiis Hauge, Belinda Cleeland and Douglas Clyde Wilson, *Fisheries Depletion and Collapse*, case study accompanying International Risk Governance Council, *Risk Governance Deficits: An Analysis and Illustration of the Most Common Deficits in Risk Governance*, Lausanne, 2009.
21 Antonello, 'The greening of Antarctica', pp. 258–59.
22 Stephen Nicol, Jacqueline Foster and So Kawaguchi, 'The fishery for Antarctic krill: recent developments', *Fish and Fisheries*, vol. 13, no. 1, March 2012, pp. 30–40.
23 Stephen Rintoul and John Church, 'The Southern Ocean's global reach', *Australian Antarctic Magazine*, no. 4, spring 2002, pp. 2–4.
24 Francis Downes Ommanney, *South Latitude*, Longmans, Green, London, 1938, p. 166.
25 Cited in James Marr, 'The natural history and geography of the Antarctic krill (*Euphausia superba* Dana)', in National Institute of Oceanography, *Discovery Reports*, vol. 32, Cambridge University Press, Cambridge, 1964, pp. 151, 210 (pp. 37–463). See also Alister C Hardy, *The Open Sea: Its Natural History*, vol. 1, *The World of Plankton*, Houghton Mifflin, Boston, 1956.
26 John Kernan, pers. comm., 11 November 2017.
27 Marr, 'The natural history and geography of the Antarctic krill'.
28 Antonello, 'The greening of Antarctica', p. 193.
29 Patti Hagan, 'The singular krill', *New York Times*, 9 March 1975, accessed 10 December 2017, <www.nytimes.com/1975/03/09/archives/the-singular-krill.html>.
30 Jacob Darwin Hamblin, *Oceanographers and the Cold War: Disciples of Marine Science*, University of Washington Press, Seattle, 2005.
31 Cassandra M Brooks and David G Ainley, 'Fishing the bottom of the Earth: the political challenges of ecosystem-based management', in Alan D Hemmings, Klaus Dodds and Peder Roberts (eds), *Handbook on the Politics of Antarctica*, Edward Elgar, Cheltenham, 2017, p. 422 (pp. 422–38).
32 Arthur Tansley, 'The use and abuse of vegetational concepts and terms', *Ecology*, vol. 16, no. 3, 1935, pp. 284–307; Frank B Golley, *A History of the Ecosystem Concept in Ecology: More than the Sum of the Parts*, Yale University Press, New Haven, 1993, p. 8.
33 Eugene P Odum, *Fundamentals of Ecology*, Saunders, Philadelphia, 1953.
34 Cited in Alessandro Antonello, 'Protecting the Southern Ocean ecosystem: the environmental protection agenda of Antarctic diplomacy and science', in Wolfram Kaiser and Jan-Henrik Meyer (eds), *International Organizations and Environmental Protection: Conservation and Globalization in the Twentieth Century*, Berghahn, New York, 2016, pp. 268–69 (pp. 268–92).
35 *United Nations Convention on the Law of the Sea*, Montego Bay, signed 10 December 1982, 1835 UNTS 397 (English), entered into force 16 November 1994, p. 397.
36 Henrik Osterblom and Olof Olsson, 'CCAMLR: an ecosystem approach to

the Southern Ocean in the Anthropocene', in Hemmings, Dodds and Roberts, *Handbook on the Politics of Antarctica*, pp. 408–409 (pp. 408–21).

37 See *Convention on the Conservation of Antarctic Marine Living Resources*, Canberra, signed 20 May 1980, 1329 UNTS 47, entered into force 7 April 1982.

38 Robert Burton, *South Georgia*, 4th edn, Government of South Georgia and the South Sandwich Islands, Stanley, 2016, p. 22.

39 For a list of the countries and dates of CCAMLR member nations, see Secretariat of the Antarctic Treaty, 'Parties', accessed 31 December 2017, <www.ats.aq/devAS/ats_parties.aspx?lang=e>.

40 The *Convention for the Conservation of Antarctic Seals*, London, signed 1 June 1972, 1080 UNTS 175, entered into force 11 March 1978, set catch limits and closed hunting seasons for the crabeater, leopard and Weddell seals and provided total protection for the much rarer Ross seal and the depleted populations of southern elephant seals and fur seals. It was the first international agreement providing for the regulation of commercial harvesting of a marine living resource before the industry had been established (or re-established). See David Ainley and Cassandra Brooks, 'Exploiting the Southern Ocean: rational use or reversion to tragedy of the commons?', in Daniela Liggett and Alan D Hemmings (eds), *Exploring Antarctic Values: Proceedings of the Workshop Exploring Linkages Between Environmental Management and Value Systems; The Case of Antarctica*, SCAR Social Science Action Group, Gateway Antarctica Special Publication Series no. 1301, University of Canterbury, Christchurch, 2013, p. 145 (pp. 143–54).

41 See Wendy Pyper, 'The good oil on krill', *Australian Antarctic Magazine*, no. 28, June 2015, pp. 1–3.

42 Andrea Thompson, 'Krill are disappearing from Antarctic waters', *Scientific American*, 29 August 2016, accessed 10 January 2018, <www.scientificamerican.com/article/krill-are-disappearing-from-antarctic-waters/>, first published as 'Climate change could put tiny krill at big risk', *Climate Central*, 26 August 2016, accessed 20 February 2018, www.climatecentral.org/news/climate-change-could-put-tiny-krill-at-risk-20641>.

43 Andrea Piñones and Alexey V Fedorov, 'Projected changes of Antarctic krill habitat by the end of the 21st century', *Geophysical Research Letters*, vol. 43, no. 16, 28 August 2016, pp. 8580–89.

44 Kristiina A Vogt, John C Gordon, John P Wargo, Daniel J Vogt, Heidi Asbjornsen, Peter A Palmiotto, Heidi J Clark, Jennifer L O'Hara, William S Keeton, Toral Patel-Weynand and Evie Witten, *Ecosystems: Balancing Science with Management*, Springer, New York, 1997, pp. 13–114.

45 For a detailed analysis of this question, see Alan D Hemmings, Klaus Dodds and Peder Roberts, 'Introduction: the politics of Antarctica', in Hemmings, Dodds and Roberts, *Handbook on the Politics of Antarctica*, pp. 1–17; Osterblom and Olsson, 'CCAMLR', pp. 416–17; Virginia Gascon and Rodolfo Werner, 'CCAMLR and Antarctic krill: ecosystem management around the great white continent', *Sustainable Development Laws & Policy*, vol. 7, no. 1, autumn 2006, pp. 14–16.

46 Frederic Briand, Theo Colborn, Richard Dawkins, Jared Diamond, Sylvia Earle, Edgardo Gomez, Roger Guillemin, Aaron Klug, Masakazu Konishi,

Jane Lubchenco, Alan MacDiarmid, Laurence Mee, Elliott Norse, Giuseppe Notarbartolo di Sciara, Gordon Orions, Roger Payne, Carl Safina, David Suzuki, John Terborgh, Edward O Wilson and George Woodwell, 'An open letter to the government of Japan on "scientific whaling"', *New York Times*, 20 May 2002; Phillip J Clapham, Per Berggren, Simon Childerhouse, Nancy A Friday, Toshio Kasuya, Laurence Kell, Karl-Hermann Kock, Silvia Manzanilla-Naim, Giuseppe Notabartolo Di Sciara, William F Perrin, Andrew J Read, Randall R Reeves, Emer Rogan, Lorenzo Rojas-Bracho, Tim D Smith, Michael Stachowitsch, Barbara L Taylor, Deborah Thiele, Paul R Wade and Robert L Brownell, 'Whaling as science', *BioScience*, vol. 53, no. 3, 1 March 2003, pp. 210–12.

47 *Whaling in the Antarctic (Australia v. Japan: New Zealand intervening) (Judgment)* [2014] ICJ Reports 226. See, for example, Paul Farrell, 'Australian court fines Japanese whaling company $1m for "intentional" breaches', *Guardian*, 18 November 2015, accessed 10 December 2017, <www.theguardian.com/environment/2015/nov/18/australian-court-fines-japanese-whaling-company-1m-for-intentional-breaches>; Daniel Flitton, 'Japan kills whale in Australian sanctuary as hunters give Sea Shepherd the slip', *Sydney Morning Herald*, 16 January 2017, accessed 10 December 2017, <www.smh.com.au/federal-politics/political-news/japan-kills-whale-in-australian-sanctuary-as-hunters-give-sea-shepherd-the-slip-20170116-gtsd4l.html>.

48 *Protocol on Environmental Protection to the Antarctic Treaty*, Madrid, signed 4 October 1991, XI ATSCM/2/3/2, reprinted in 30 ILM 1461, entered into force 1998; Committee for Environmental Protection, *25 Years of the* Protocol on Environmental Protection to the Antarctic Treaty, Secretariat of the Antarctic Treaty, 2016, accessed 11 January 2018, <www.ats.aq/documents/atcm39/ww/atcm39_ww007_e.pdf>.

49 Commission for the Conservation of Antarctic Marine Living Resources, 'Toothfish fisheries', last modified 22 March 2017, accessed 6 October 2017, <www.ccamlr.org/en/fisheries/toothfish-fisheries>; Orea RJ Anderson, Cleo J Small, John P Croxall, Euan K Dunn, Benedict J Sullivan, Oliver Yates and Andrew Black, 'Global seabird bycatch in longline fisheries', *Endangered Species Research*, vol. 14, 2011, pp. 91–106; Stephen Leahy, 'The UN starts a conservation treaty for the high seas', *National Geographic*, 24 December 2017, accessed 30 December 2017, <news.nationalgeographic.com/2017/12/un-high-seas-conservation-treaty-ocean-protection-spd/>.

50 Wallace S Broecker, *The Great Ocean Conveyor: Discovering the Trigger for Abrupt Climate Change*, Princeton University Press, Princeton, 2010.

51 Susan Barr and Cornelia Lüdecke (eds), *The History of the International Polar Years (IPYs)*, Springer-Verlag, Berlin, 2010.

52 'Southern Ocean winds open window to the deep sea', *Siroscope*, Commonwealth Science and Industrial Research Organisation, no. 61, March 2010, accessed 14 July 2016, <www.csiro.au/news/newsletters/SIROscope/2010/March10/htm/deepsea.htm>.

53 Stephen Rintoul, 'Seeing under the ice: a strategy for observing the Southern

Ocean beneath sea ice and ice shelves', introduction talk at Seeing Below the Ice workshop, Hobart, 22–25 October 2012.

54 Nathan Bindoff, Stephen Rintoul and Marcus Haward, *Position Analysis: Climate Change and the Southern Ocean*, Antarctic Climate and Ecosystems Cooperative Research Centre, Hobart, 2011, pp. 15, 20.

55 See, for example, Stephen Rintoul, Nathan Bindoff, Will Hobbs, Beatriz Peña-Molino, Stephanie Downes and Mark Rosenberg, 'The Southern Ocean in a changing climate', Project R1.1, Antarctic Climate and Ecosystems Cooperative Research Centre, accessed 19 January 2018, <acecrc.org.au/project/southern-ocean-in-a-changing-climate/>; Robert Kunzig, *Mapping the Deep: The Extraordinary Story of Ocean Science*, WW Norton, New York, 2000, p. 291.

56 Rintoul and Church, 'The Southern Ocean's global reach'.

57 Stephen Rintoul, Wenju Cai, Helen Cleugh and Gongke Tan, 'Our new research centre focuses on the "ocean hemisphere"', *CSIROscope*, 22 May 2017, accessed 15 July 2017, <blog.csiro.au/new-research-centre-focuses-on-the-ocean-hemisphere/>.

58 Tony Press, 'In deep: Australian research in the Southern Ocean', *Australian Antarctic Magazine*, no. 4, spring 2002, pp. 1–2.

59 See, for example, Rintoul et al., 'The Southern Ocean in a changing climate'; Nisha Harris, 'Antarctic Bottom Water disappearing', *Australian Antarctic Magazine*, no. 23, December 2012, p. 18.

60 Bindoff, Rintoul and Haward, *Position Analysis*, pp. 3–4; Australian Antarctic Division, 'Ocean acidification', Department of the Environment and Energy (Australia), last updated 28 September 2007, accessed 14 November 2017, <antarctica.gov.au/about-antarctica/environment/climate-change/ocean-acidification-and-the-southern-ocean>.

61 Commonwealth Science and Industrial Research Organisation, 'Oceans and climate change', last updated 23 March 2015, accessed 21 June 2016, <www.csiro.au/en/Research/Environment/Oceans-and-coasts/Oceans-climate?ref=/CSIRO/Website/Research/Environment/Atmosphere-and-climate/Oceans-climate>; Australian Antarctic Division, 'Southern Ocean ecosystems', Department of the Environment and Energy (Australia), last updated 27 October 2015, accessed 2 November 2017, <antarctica.gov.au/science/climate-processes-and-change/marine-ecosystem-change>.

62 Bindoff, Rintoul and Haward, *Position Analysis*, p. 3.

63 Matthew England, evidence presented at a public hearing of the Paris Agreement inquiry, Sydney, 27 September 2016, transcript: Joint Standing Committee on Treaties, *Paris Agreement*, Commonwealth of Australia, Canberra, 27 September 2016, pp. 15–17 (pp. 15–21).

64 Bindoff, Rintoul and Haward, *Position Analysis*, p. 17.

65 David Barnes, 'Antarctic Seabed Carbon Capture Change', British Antarctic Survey, accessed 19 February 2018, <www.bas.ac.uk/project/antarcticseabedcarboncapturechange/>.

66 Elisabeth Mann Borgese, *The Oceanic Circle: Governing the Seas as a Global*

Resource, United Nations University Press, Tokyo, 1998; Antonello, 'The greening of Antarctica, p. 229.

67 Herman Melville, *Moby-Dick; or, The Whale*, R Bentley, London, 1851; Jules Verne, *Vingt mille lieues sous les mers: tour du monde sous-marin* [Twenty Thousand Leagues Under the Sea: A Tour of the Underwater World], Hetzel, Paris, 1871 (serialised 1869–70).

68 Rachel Carson, *The Sense of Wonder*, Harper & Row, San Francisco, 1956; *Blue Planet II*, television documentary series, Natural History Unit, BBC, London, 2017.

69 Anthropocene is a recent term used to describe Earth's present geological epoch as one in which humans have made lasting impacts on the physical environment and climate. See, for example, Osterblom and Olsson, 'CCAMLR', pp. 408–21.

70 Amitav Ghosh, 'Where is the fiction about climate change?', *Guardian*, 28 October 2016, accessed 15 December 2017, <www.theguardian.com/books/2016/oct/28/amitav-ghosh-where-is-the-fiction-about-climate-change->; Amitav Ghosh, *The Great Derangement: Climate Change and the Unthinkable*, University of Chicago Press, Chicago, 2016, p. 8.

71 Elizabeth Leane and Stephen Nicol, 'Charismatic krill? Size and conservation in the ocean', *Anthrozoös*, vol. 24, no. 2, 2011, pp. 135–46.

72 Press, 'In deep'.

73 Translated from the Chilean-language inscription.

SELECT BIBLIOGRAPHY

Amundsen, Roald, *The South Pole: An Account of the Norwegian Antarctic Expedition in the* Fram, *1910–1912*, J Murray, London, 1912.

Anson, George, *A Voyage Round the World in the Years MDCCXL, I, II, III, IV*, eds Richard Walter and Benjamin Robins, Open University Press, New York, 1974 (1749).

Barr, Susan and Cornelia Lüdecke (eds), *The History of the International Polar Years (IPYs)*, Springer-Verlag, Berlin, 2010.

Beaglehole, John Cawte, *The Life of Captain James Cook*, Stanford University Press, Stanford, 1974.

Bellingshausen, Fabian Gottlieb Thaddeus von, *The Voyage of Captain Bellingshausen to the Antarctic Seas, 1819–1821*, 2 vols, ed. Frank Debenham, trans. from Russian, Hakluyt Society, London, 1945.

Bennett, Isobel, *Shores of Macquarie Island*, Rigby, Adelaide, 1971.

Bernacchi, Louis, *To the South Polar Regions: Expedition of 1898–1900*, Bluntisham, Denton, 1991 (1901).

Broecker, Wallace S, *The Great Ocean Conveyor: Discovering the Trigger for Abrupt Climate Change*, Princeton University Press, Princeton, 2010.

Bruun, Anton F, Svend Greve, Hakon Mielche and Ragnar Spärck (eds), *The* Galathea *Deep Sea Expedition, 1950–1952, Described by Members of the Expedition*, trans. Reginald Spink, G Allen & Unwin, London, 1956.

Bull, Henrik Johan, *The Cruise of the 'Antarctic' to the South Polar Regions*, E Arnold, London, 1896.

Carson, Rachel, *The Sea Around Us*, Oxford University Press, Oxford, 1951.

Chapman, Anne, *European Encounters with the Yámana People of Cape Horn, Before and After Darwin*, Cambridge University Press, Cambridge, 2010.

Cherry-Garrard, Apsley, *The Worst Journey in the World*, Vintage, London, 2010 (1922).

Coleman-Cooke, John, Discovery II *in the Antarctic*, Odhams, London, 1963.

Cook, James, *A Voyage Towards the South Pole, and Round the World: Performed in His Majesty's Ships the* Resolution *and* Adventure, *in the Years 1772, 1773, 1774, and 1775*, 2 vols, W Strahan & T Cadell, London, 1777.

Corbin, Alain, *The Lure of the Sea: The Discovery of the Seaside in the Western World, 1750–1840*, trans. Jocelyn Phelps, University of California Press, Berkeley, 1994.

Corfield, Richard, *The Silent Landscape: The Scientific Voyage of HMS* Challenger, Joseph Henry, Washington, DC, 2003.

Cottee, Kay, *First Lady: A History-making Solo Voyage Around the World*, Macmillan, South Melbourne, 1989.
Dalrymple, Alexander, *A Collection of Voyages Chiefly in the Southern Atlantick Ocean*, J Nourse, London, 1775.
Dalton, John N (ed.), *The Cruise of H.M.S. 'Bacchante', 1879–1882*, Macmillan, London, 1886.
Dampier, William, *A New Voyage Round the World*, J Knapton, London, 1697.
Darwin, Charles, *Journal and Remarks, 1832–1836*, vol. 3 of Robert FitzRoy (ed.), *Narrative of the Surveying Voyages of His Majesty's Ships* Adventure *and* Beagle, *Between the Years 1826 and 1836, Describing Their Examination of the Southern Shores of South America, and the* Beagle*'s Circumnavigation of the Globe*, H Colburn, London, 1839.
Deacon, George ER, *The Antarctic Circumpolar Ocean*, Cambridge University Press, Cambridge, 1984.
Fogg, Gordon Elliott, *A History of Antarctic Science*, Cambridge University Press, Cambridge, 1992.
Forster, Georg, *A Voyage Round the World in His Britannic Majesty's Sloop,* Resolution, *Commanded by Capt. James Cook, During the Years 1772, 3, 4, and 5*, 2 vols, B White, etc., London, 1777.
Forster, Johann Reinhold, *Observations Made During a Voyage Round the World*, eds Nicholas Thomas, Harriet Guest and Michael Dettelbach, University of Hawaii Press, Honolulu, 1996 (1778).
Garrison, Tom, *Oceanography: An Invitation to Marine Science*, Brooks / Cole-Thomson Learning, Belmont, 2005.
Gillham, Mary E, *Sub-antarctic Sanctuary: Summertime on Macquarie Island*, AH & AW Reed, Wellington, 1967.
Gregory, John Walter, *The Climate of Australasia in Reference to Its Control by the Southern Ocean*, Whitcombe & Tombs, Melbourne, 1904.
Griffiths, Tom, *Slicing the Silence: Voyaging to Antarctica*, NewSouth, Sydney, 2007.
Hamblin, Jacob Darwin, *Oceanographers and the Cold War: Disciples of Marine Science*, University of Washington Press, Seattle, 2005.
Hardy, Alister C, *Great Waters: A Voyage of Natural History*, Harper & Row, New York, 1967.
Heezen, Bruce C and Charles D Hollister, *The Face of the Deep*, Oxford University Press, New York, 1971.
Hemmings, Alan D, Klaus Dodds and Peder Roberts (eds), *Handbook on the Politics of Antarctica*, Edward Elgar, Cheltenham, 2017.
Hince, Bernadette, *The Antarctic Dictionary: A Complete Guide to Antarctic English*, CSIRO, Canberra, 2000.
Hoare, Philip, *The Sea Inside*, Melville House, New York, 2014.
Hooker, Joseph Dalton, *The Botany of the Antarctic Voyage of H.M. Discovery Ships* Erebus *and* Terror *in the Years 1839–1843, under the Command of Captain Sir James Clark Ross, Kt., R.N., F.R.S. &c.*, 6 vols, Reeve, London, 1844–59.
Knox, George A, *Biology of the Southern Ocean*, 2nd edn, CRC, Boca Raton, 2007.
Kunzig, Robert, *Mapping the Deep: The Extraordinary Story of Ocean Science*, WW Norton, New York, 2000.

Larson, Edward J, *An Empire of Ice: Scott, Shackleton, and the Heroic Age of Antarctic Science*, Yale University Press, New Haven, 2011.
Lyell, Charles, *Principles of Geology, Being an Attempt to Explain the Former Changes of the Earth's Surface, by Reference to Causes Now in Operation*, 3 vols, J Murray, London, 1830–3.
Markham, Clements R, *The Lands of Silence*, Cambridge University Press, Cambridge, 1921.
Matkin, Joseph, *At Sea with the Scientifics: The* Challenger *Letters of Joseph Matkin*, ed. Philip F Rehbock, University of Hawaii Press, Honolulu, 1992.
Maury, Matthew Fontaine, *The Physical Geography of the Sea*, 3rd edn, Harper, New York, 1855.
Mawson, Douglas, *Geographical Narrative and Cartography*, Australian Antarctic Expedition 1911–14, Scientific Reports, series A, vol. 1, Government Printer, Sydney, 1942.
Mills, Eric L, *The Fluid Envelope of Our Planet: How the Study of Ocean Currents Became a Science*, University of Toronto, Toronto, 2009.
Moseley, Henry Nottidge, *Notes by a Naturalist on the 'Challenger', Being an Account of Various Observations Made During the Voyage of H.M.S. 'Challenger' Round the World, in the Years 1872–1876, under the Commands of Capt. Sir G. S. Nares, R.N., K.C.B., F.R.S., and Capt. F. T. Thomson, R.N.*, Macmillan, London, 1879.
Murphy, Robert Cushman, *Logbook for Grace: Whaling Brig Daily, 1912–1913*, R Hale, London, 1948.
Murray, George (ed.), *The Antarctic Manual for the Use of the Expedition of 1901*, Royal Geographical Society, London, 1901.
Murray, John and Johan Hjort, *Depths of the Ocean*, Macmillan, London, 1912.
Ommanney, Francis Downes, *South Latitude*, Longmans, Green, London, 1938.
Owen, Russell, *The Antarctic Ocean*, Whittlesey House, New York, 1941.
Ponting, Herbert, *The Great White South; or, With Scott in the Antarctic, Being an Account of Experiences with Captain Scott's South Pole Expedition and of the Natural Life of the Antarctic*, Duckworth, London, 1935 (1921).
Pyne, Stephen J, *The Ice: A Journey to Antarctica*, University of Washington Press, Seattle, 1986.
Raban, Jonathan, *Passage to Juneau: A Sea and Its Meanings*, Picador, Surrey, 1999.
Ross, James Clark, *A Voyage of Discovery and Research in the Southern and Antarctic Regions, During the Years 1839–43*, 2 vols, J Murray, London, 1847.
Rozwadowski, Helen, *Fathoming the Ocean: The Discovery and Exploration of the Deep Sea*, Belknap Press of Harvard University Press, Cambridge, 2009.
Russell, Lynette, *Roving Mariners: Australian Aboriginal Whalers and Sealers in the Southern Oceans, 1790–1870*, State University of New York Press, Albany, 2012.
Sayed, Sayed Z El- (ed.), *Southern Ocean Ecology: The BIOMASS Perspective*, Cambridge University Press, Cambridge, 1994.
Scholes, Arthur, *Fourteen Men: The Story of the Australian Antarctic Expedition to Heard Island*, G Allen & Unwin, London, 1951.
Scott, Robert Falcon, *Scott's Last Expedition: Diaries, 26 November 1910 – 29 March 1912*, Amberley, Stroud, 2012.

Shackleton, Ernest, *South: The Story of Shackleton's Last Expedition, 1914–17*, Heinemann, London, 1919.
Slocum, Joshua, *The Voyages of Joshua Slocum*, ed. Walter Magnes Teller, Rutgers University Press, New Brunswick, 1958.
Stommel, Henry M, *Science of the Seven Seas*, Cornell Maritime Press, New York, 1945.
Swire, Herbert, *The Voyage of the* Challenger*: A Personal Narrative of the Historic Circumnavigation of the Globe in the Years 1872–1876*, Golden Cockerel, London, 1938.
Tizard, Thomas Henry, Henry Nottidge Moseley, John Young Buchanan and John Murray, *Narrative of the Cruise of H.M.S.* Challenger*, with a General Account of the Scientific Results of the Expedition*, 2 vols, 1885, part of Great Britain, Challenger Office, *Report on the Scientific Results of the Voyage of H.M.S.* Challenger *During the Years 1873–76 under the Command of Captain George S. Nares, R.N., F.R.S. and [the Late] Captain Frank Tourle Thomson, R.N.*, 50 vols, Neill, Edinburgh, 1880–95.
Webster, William Henry Bayley, *Narrative of a Voyage to the Southern Atlantic Ocean, in the Years 1828, 29, 30, Performed in H.M. Sloop* Chanticleer*, under the Command of the Late Captain Henry Foster, F.R.S. &c.*, R Bentley, London, 1834.
Wilkes, Charles, *Narrative of the United States Exploring Expedition During the Years 1838, 1839, 1840, 1841, 1842*, 5 vols, Lea & Blanchard, Philadelphia, 1845.
Williams, Glyn, *Naturalists at Sea: Scientific Travellers from Dampier to Darwin*, Yale University Press, New Haven, 2013.
Wilson, Edward Adrian, *Diary of the* Discovery *Expedition to the Antarctic Regions, 1901–1904*, ed. Ann Savours, Blandford, London, 1966.
Worsley, Frank Arthur, Endurance*: An Epic of Polar Adventure*, P Allan, London, 1931.
Wu, Norbert and Jim Mastro, *Under Antarctic Ice: The Photographs of Norbert Wu*, University of California Press, Berkeley, 2004.

INDEX

Please note: numbers in italics refer to numbered photographs.

Index

Index

Index

Index

Index